Advanced Remote Sensing Technology for Synthetic Aperture Radar Applications, Tsunami Disasters, and Infrastructure

Edited by Maged Marghany

Published in London, United Kingdom

IntechOpen

Supporting open minds since 2005

Advanced Remote Sensing Technology for Synthetic Aperture Radar Applications, Tsunami Disasters, and Infrastructure
http://dx.doi.org/10.5772/intechopen.78525
Edited by Maged Marghany

Contributors
Chung Fu, Yifan Zhu, Kuangyuan Hou, Chang Luo, Hanqiao Huang, Yong Wang, Shiqiang Wang, Dong Li, Yunhua Zhang, Xiaojin Shi, John Trinder, Qingxiang Liu, Yoichi Kunii, Ryo Natsuaki, Abu Sambah, Fusanori Miura, Maged Marghany

First published in London, United Kingdom, 2019 by IntechOpen
IntechOpen is the global imprint of INTECHOPEN LIMITED, registered in England and Wales, registration number: 11086078, The Shard, 25th floor, 32 London Bridge Street
London, SE19SG - United Kingdom
Printed in Croatia

British Library Cataloguing-in-Publication Data
A catalogue record for this book is available from the British Library

Additional hard copies can be obtained from orders@intechopen.com

Advanced Remote Sensing Technology for Synthetic Aperture Radar Applications, Tsunami Disasters, and Infrastructure
Edited by Maged Marghany
p. cm.
Print ISBN 978-1-78985-617-0
Online ISBN 978-1-78985-618-7

For EU product safety concerns:
IN TECH d.o.o., Prolaz Marije Krucifikse Kozulić 3, 51000 Rijeka, Croatia, info@intechopen.com or visit our website at intechopen.com.

Meet the editor

Prof. Dr. Maged Marghany is the author of "Advanced Remote Sensing Technology for Tsunami Modelling and Forecasting" which was published by Routledge Taylor and Francis Group, CRC. Prof. Dr. Maged Marghany is currently a professor of microwave remote sensing at Geomatika University College, Malaysia. He has led several projects related to the application of SAR to Malaysian coastal waters, funded by the Ministry of Science and Technology, Malaysia (MOSTE) and also the Ministry of High Education in Malaysia (MOHE). His research is directed towards the use of synthetic aperture radar (SAR) data for modeling shoreline changes and developing a new approach for forecasting oil slick trajectory movements. He teaches at postgraduate levels on the topic of digital image processing and microwave remote sensing.
Prof. Maged Marghany established a new research method regarding four-dimensional forecasting, which was published in his book "Advanced Remote Sensing Technology for Tsunami Modelling and Forecasting" in addition to involving quantum theory for forecasting tsunamis.
Prof. Dr. Maged Marghany (1967) was awarded the ESA Post-doctoral Fellowship by the International Institute of Aerospace and Earth Observation (ITC) in Enschede, the Netherlands, funded by the European Space Agency (ESA) for one year from March 2000 to March 2001. Prof. Dr. Maged has published over 270 papers in international conferences and indexed journals. He is also the editor of three books published by the rapidly growing IntechOpen publisher.

Contents

Preface

The advance in space machineries has created a novel technology for observing and monitoring the Earth from space. Most earth observation remote sensing considerations focus on using conventional image processing algorithms or classic edge detection tools. Nevertheless, these techniques do not implement modern physics, applied mathematics, signal communication, remote sensing data, and innovative space technologies. This book provides readers with methods to comprehend how to monitor coastal environments, disaster areas, and infrastructure from space with advanced talent remote sensing technology to bridge the gaps between modern space technology, image processing algorithms, mathematical models and the critical issue of the coastal and infrastructure investigations. In other words, advanced remote sensing technology, which covers sensor developments, and image processing algorithm modifications, which are based on modern physics, artificial intelligence, and machine learning. In these regards, their applications cover a wide range of coastal observations, for instances, high risk of a tsunami depends on the depth of water, the coastal geomorphology, the direction of the tsunami wave, and the existence of rivers or other water canals. In these circumstances, coastal zones are required for new urban planning and specific infrastructure designing to reduce the impact of such a disaster.

In spite of numerous of synthetic aperture radar (SAR) space technology, the developing country researchers and scientists are still focusing on optical remote sensing technology. In fact, microwave remote sensing require use of mathematics and physics behind the SAR technology. The first chapter introduces a new technology for measuring sea surface current using along-track interferometry of TanDEM-X satellite data. This chapter delivers a novel algorithm to retrieve sea surface current using the multichannel MAP height estimator algorithm, which is considered the first study of ocean current in the coastal waters of Peninsular Malaysia.

The available SAR data increases dramatically with the recent operation of many spaceborne and airborne SAR systems. This makes the joint processing of multiple images for accurate understanding and perception of a scene and target possible. For SAR image pairs acquired from different imaging geometries or by different sensors, there is always a geometrical warp between them, which should be compensated first before any deep application. Image registration is aimed to retrieve the warp function to align the same pixel position in each SAR image to the same target position in the global system. A lot of SAR image registration techniques have been developed hitherto. In the second chapter, the algorithms that conduct registration based on image features, such as contour, region, line, and point are accurately addressed. Contour, region, and line, as well as their combinations, are often used for registration of multi-modality images. For SAR images with geometrical distortion and speckle, point feature is generally much clearer and easier extracted. Tie points, corner, and key points are the commonly-used features in SAR image registration. Tie points usually refer to the features extracted from tie patches in SAR image registration.

Following the second chapter, the third chapter introduces a novel technology for implementation of interferometry synthetic aperture radar (InSAR) with L-band on monitoring harbor. In fact, L-band SAR and its long-lasting temporal coherence is an advantage to perform precise interferometric coherence analysis. In addition, recent high-resolution SAR images are found to be useful for observing relatively small targets, e.g., individual buildings and facilities. In this chapter, author presents the basic theory of SAR observation, interferometric coherence analysis for the disaster monitoring and its examples of the harbor facilities.

However, optical remote sensing experts are relying on commercial software and open source codes without fully understanding the mathematical algorithms involved in image processing. In fact, conventional image processing techniques such as image classification are being used. In this view, classifying remote scenes according to a set of semantic categories is a very challenging problem, because of high intraclass variability and low interclass distance.

The most advanced image processing technique is presented in Chapter 4. One of the advanced learning machine algorithms for image processing is Deep convolutional neural networks (CNNs), which have been widely used to obtain high-level representation in various computer vision tasks. Deep CNN models are trained upon a database of more than 1.2 million categorized natural images of 1000+ classes, which serve as the backbone for many segmentation, detection and classification tasks on other data sets.

The fifth chapter presents an image processing technique that is based on the sub-pixel algorithms for modeling a time series of shoreline changes. In fact, the majority of investigations are only used conventional classification or threshold technique to study short periods of coastal erosion. The novelty of the fifth chapter is that the authors implemented eight years of time series of multispectral data with a subpixel technique to reduce the error of shoreline extraction at sub-pixel, pixel and object-based scales.

A different remote sensing technique is introduced in the sixth chapter for monitoring infrastructure. This technique is based on utilization of accelerometer measurements. In fact, infrastructure, including roads, bridges, tunnels, water supply, sewers, electrical grids, and telecommunications, may be exposed to environmentally-induced or traffic-induced vibrations. Some infrastructure, such as bridges and roadside upright structures, may be sensitive to vibration where accelerometers and other types of sensors may be used for their measurement of sensitivity to environmentally-induced loads, such as wind and earthquakes, and traffic-induced loads, such as passing trucks. With data collected by accelerometers, time histories may be obtained, transformed, and then analyzed to determine their modal frequencies and shapes.

A coastal disaster, which is mainly based on a tsunami disaster, requires such advanced technology of geospatial monitoring of the tsunami risk impact. The seventh chapter delivers the integrated approach of raster weighted overlay of all spatial databases of tsunami vulnerability and risk parameters specifying the vulnerability and risk area due to the tsunami and defines the possible area that could be affected by the tsunami and the potential inundated area.

Finally, the book describes a new technology of Unmanned Aerial Vehicle for accurate three-dimensional reconstruction (3-D). In fact, this technology is rapidly growing among the researchers and scientists. Chapter eight presents a novel technology to construct a precise 3-D image using an Unmanned Aerial Vehicle, which is validated by ground field measurements during UAV experiments. In this view, there is great potential to use UAV images for 3-D modeling when compared to operational satellite data.

I wish to convey my appreciation to all authors who contributed novel work to this book. Without their intense commitment, this book would not have become such a precious piece of novel knowledge. I am also grateful to the IntechOpen editorial team Ms. Martina Josavac and Ms. Maja Bozicevic who afforded the opportunity to publish this book.

Prof. Dr. Maged Marghany
Microwave Remote Sensing expert,
Faculty Geospatial and Real Estate,
Geomatika University College,
Kuala Lumpur, WP Kuala Lumpur,
Malaysia

Section 1

Synthetic Aperture Radar Image Processing and Applications

Chapter 1

Introductory Chapter: Advanced Ocean Current Simulation from TanDEM Satellite Data

Maged Marghany

1. Introduction

Satellite microwave data, such as synthetic aperture radar (SAR), have the great potential for retrieving ocean dynamic parameters, for instance, ocean surface current and ocean wave dynamic [1]. One of the attention-grabbing topics is current flow that is needed for short go back satellite cycle and high resolution. These will provide precisely data concerning current dynamic flow [2, 3]. In fact, current is very important for ship navigation, fishing, waste matter substances transport, and sediment transport [4, 5]. Respectively, optical and microwave sensors are enforced to monitor the current flows. Indeed, the ocean surface dynamic options of sea surface current are vital parameters for atmospheric-sea surface interactions. In this regard, the global climate change, marine pollution, and coastal risky are preponderantly dominated by current speed and direction [1]. The measurements of ocean current from space rely on the electromagnetic signal. Truly, associate degree of an electromagnetic signal of optical and microwave reflects from the ocean carrying records concerning one among the first discernible quantities that are the color, the beamy temperature, the roughness, and also the height of the ocean [2].

Recently, the high resolution of SAR sensors such as TerraSar-X, RADARSAT-2, ALOS PALSAR, and the foremost three of the Italian satellite of COSMO-SkyMed have been commenced. Once the four satellites in the COSMO-SkyMed constellation are developed, they are conceivable functioning with a tiny resume time of a little hours [4]. Nevertheless, the initial three of the COSMO-SkyMed, ALOS PALSAR, and RADARSAT-2, satellite data are the cross-track interferometry, which do not allow determining neither coastal water flow nor coastal water level changing. In this regard, the TerraSAR-X satellite data use an along-track interferometric proficiency which simply permits the quantity of sea surface speed. Additionally, phase alterations between the coregistered pixels of an image pair are consistent to Doppler frequency shifts of the signal backscattered and according to line-of-sight velocities of the scatterers. In this view, phase alterations include influences of surface flows and of the dynamic of wave movement. Consequently, the retrieving of tidal current flow can be accurately achieved by both of TerraSAR-X and TanDEM-X. These can be depleted to regulate precisely coastal water height fluctuations. The TerraSAR-X can regulate perfectly the digital surface model (DSM), where depiction of surface-containing topographies exceeds the terrain height, for example, plants and constructions through precision of 2 m.

Moreover, TanDEM-X involves dual high-resolution imaging SAR data. In this understanding, both TerraSAR-X and TanDEM-X are hovering in tandem and establishing an enormous radar interferometer with an anticipated competence of creating a comprehensive DSM through a perpendicular resolution of 2 m, exceeding whatever obtainable currently from space [4]. Consistent with Romeiser et al. [6], with the usual helical revolution configuration, the dual satellites ensure an along-track gap between 0 at the northern and southern utmost locations of the orbit and approximately 550 m over the equator, restrictive of the district of convenient baselines for intersatellite interferometry above the sea surface to restricted space crews far-off the north and south. In districts of elongated along-track baselines, the data characteristic undergoes since sequential decorrelation of the signal backscattered. Nonetheless, the TanDEM-X geometry constructions acquire adjusted from period to period to enhance the cross-track interferometry performing in coastal water height fluctuations and surface stream flow attentions [7].

On the word of Yoon et al. [8], the phase computation is a foremost encounter to regulate surplus precise height. This is because the calculated phase differences are assumed as a wrapped phase of the primary quantities of a scale $-\pi$ to π, hence the actuality vague contained by multiples of 2π [2, 9, 10]. This technique generates phase leaps between nearby pixels. Smooth function is depleted to resolve phase leap through adding or detracting multiples of 2π. Subsequently, Ferraiuolo et al. [2] have developed the multichannel MAP height estimator as a function of a Gaussian Markov random (GMRF) to unravel the doubts of height retrieving from InSAR procedure. They initiated that the multichannel MAP height estimator has accomplished the phase gaps and tweaked the height contour as compared to predictable phase unwrapping set of rules, i.e., path-following algorithms and minimum-norm algorithms.

The foremost demonstrable of this experiment is to investigate the coastal water level and velocity changes using along-track interferometric synthetic aperture radar (ATInSAR) technique multichannel MAP height estimator.

2. Algorithm

The algorithm is implemented in this study, which is based on the multichannel MAP height estimator. It is depleted to retain the information of the sea surface level alterations. This algorithm is implemented from the consideration of Baselice et al. [9]. Succeeding Baselice et al. [9], the signal of interferometric phase can be articulated by the next mathematical Eq. (1) [9],

$$\phi_{sn} = \left\langle \left(\frac{4\pi}{\lambda R_0 \sin\theta} \right) B_{\perp n} h_s + \alpha \right\rangle_{2\pi}, \ \mathrm{n} = 1, 2, \ldots, \mathrm{N}; s = 1, 2, \ldots, \mathrm{S} \tag{1}$$

where s is the pixel locus in the TanDEM-X data, n is the deliberated interferogram band, λ is the TanDEM-X wavelength, R_0 is the stretch between the epicenter of the sight and the controlling antenna, and $B_{\perp n}$ is the orthogonal baseline. Furthermore, h_s is the height rate in meter, α is the phase decorrelation noise, and an incident angle is presented by θ. Moreover, $\langle . \rangle_{2\pi}$ signifies the "modulo-2π." Let us assume that N is autonomous interferogram bands; thenceforth, the obstruction contains the retrieving of the sea-level height rates h_s, which is being from the $S \times N$ as a function of the expected wrapped phase ϕ_{sn}. Succeeding Ferraiuolo et al. [2], the obstruction of demonstrating height can be elucidated by means of a MAP height approximation technique. In this understanding, the multichannel probability function F_{mc} is formulated as:

$$F_{mc}(\phi_s|\varsigma_s) = \prod_{n=1}^{N} f((\phi_{sn}|\varsigma_s) \tag{2}$$

here $F(\phi_{sn}|\varsigma_s)$ is the likelihood function of the signal channel, ϕ_s is calculated as wrapped phase data which is denoted as the pixel s, $\phi_s = [\phi_{s1}, \phi_{s2}, \ldots\ldots, \phi_{sN}]^T$, and ς_s

is collected vector height values where $\varsigma = [\varsigma_1, \varsigma_2, \ldots\ldots, \varsigma_S]^T$. Succeeding Baselice et al. [9] and Ferraiuolo et al. [2], a MAP algorithm height approximation can be casted by:

$$\widehat{\varsigma_{MAP}} = \arg_{\varsigma}\max\ln\left[\left(\prod_{S=1}^{S} F_{mc}(\phi_s|\varsigma_s)g(\varsigma;\hat{\sigma}\right)\right] \tag{3}$$

here $g(.)$ is a preceding probability density function (pdf) which is approved by means of Gaussian Markov random field (GMRF) and $\hat{\sigma}$ is the hyperparameter route which is not a preceding identified. As said by Baselice et al. [9], GMRF can be appraised beginning from the restrained interferograms. This is realized by deliberating subbands, equivalent to diverse azimuth looks. In this regard, GMRF is determined by:

$$g(\varsigma, \hat{\sigma}) = \frac{1}{Z(\sigma)} e^{\left(-\sum_{S=1}^{SxN}\sum_{K\in N_S}\left[\frac{(\varsigma_s-\varsigma_k)^2}{2\sigma_{sk}^2}\right]\right.} \tag{4}$$

where N_s is the district system of sth pixel, and s is well known as hyperparameters, which are illustrative of the confined physical appearance of the sea-level height h, σ is the hyperparameter vector assembling all pixel values, and Z(σ) is the detachment function [10] which is required to standardize the pdf [2, 9, 10]. Lastly, the regularized restoration square error is estimated via:

$$\varepsilon = \frac{||\hat{\varsigma} - \varsigma||^2}{||\varsigma||^2} \tag{5}$$

where ς is the sea-level height which is derived from Eq. (3) and the accurate height then can be estimated from Eq. (4) ($\hat{\varsigma}$). Though the reform is deliberating the restricted sum of accessible data (four bands), it is virtuous to recover its feature, predominantly on the disjointedness. Formerly, inverse algorithm is executed to repossess the sea surface current pattern.

3. Dataset

Two panaches of acquaintance are required to inverse the sea surface current pattern which are: (i) TanDEM-X of SAR; and (ii) real in situ measurements throughout TanDEM-X satellite overpassed.

3.1 Satellite TanDEM-X data

Pair of Terra-SAR satellite data is attained by the TanDEM-X satellite on May 6, 2017. The earliest date was attained at 7:27:17 am; however, the subsequent data obtained at 19:20:06 pm. Both data are in spotlight mode with X-band and HH and VV polarization, respectively. Both spotlight modes are formatted in single look complex binary data. The TanDEM-X functioning concern encompasses the synchronized maneuver of two satellites hovering in contiguous pattern. The modification restraints for the construction are: (i) the revolution arising nodes, (ii) the perspective between the perigees, (iii) the revolution peculiarities, and (iv) the phasing between the satellites. The adherence of ocean surface flow is a vigorous façade of evaluating climate variations. Space-borne SAR along-track interferometry (ATI) obligates the talent to greatly subsidize to the contemporary field.
It will recommend a great-area, global-widespread seeming surface flow quantities.

The difficulties of representing comparatively low-slung speeds are regular resolute by the developments of SAR satellites that produce satisfactorily considerate ATI quantities [7].

In this revision, the multichannel MAP height estimator relies on the TanDEMX facts. Both TerraSAR-X and TanDEM-X satellites transmit identical SAR devices functioning at 9.65 GHz frequency (X-band). All over approximately dedicated maneuvers, both satellites are positioned acquaintance exceptionally in an actual singular track conformation through a fleeting along path reference line delivering a possibility for sea surface flow quantities. The TanDEM-X data exploited in this investigation were bistatic (TS-X active/TD-X passive) channel with VV polarization and in stripmap (SM) [6, 7].

3.2 In situ ocean current measurement

Succeeding Marghany [3], the device of Aquadopp® 2 MHz current meter was used to acquire the physical information of sea surface flows, for instance, speed and direction (**Figure 1**). In this view, the surface flow information achievement

Figure 1.
Deployment of Aquadop 2 MHz current meter in the coastal water.

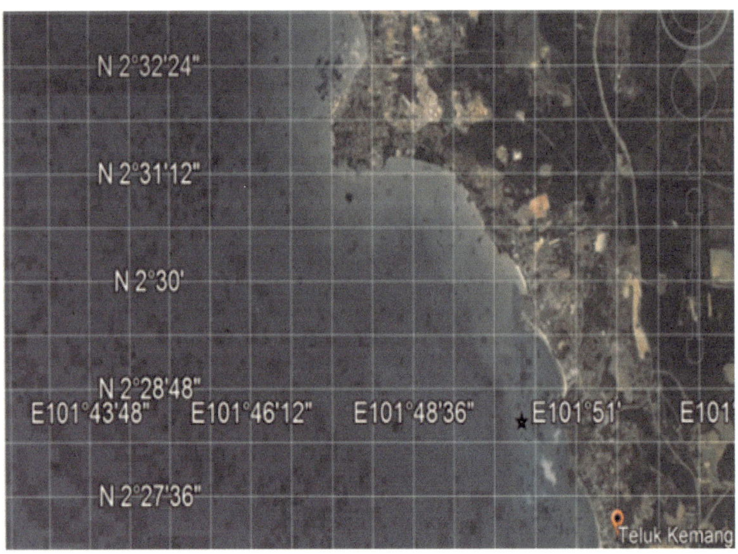

Figure 2.
In situ measurements geographical location.

was collected by the Aquadopp® 2 MHz current meter factory-made by Nortek AS, Scandinavian country. The device could be a standalone composition, manipulation of the Doppler-established frequency equipment to gauge the surface current flows at the positioning of a fixed geographical location on the sea surface. The equipment is envisioned basically with memory and internal battery pack somewhere it may be intended to tape and collect information within for self-positioning [3].

Along the coastal water of Teluk Kemang, Port Dickson, Malaysia, the current meter instrument of Aquadopp® 2 MHz current meter was arrayed on May 6, 2017 (**Figure 2**). Two periods of data collection were carried out: (i) at 6:15 am to 8:15 am and (ii) at 6:15 pm to 8:15 pm. For both phases, therefore, the surface flows were deliberated for intermissions of 2 h.

4. Current pattern from TanDEM-X data

The TanDEM-X satellite data with the spotlight of VV polarization are implemented to retrieve the sea surface flow rates (**Figure 3**). The retrieving sea level and sea surface flow variations are constrained to range direction. In fact, the sea surface current is only sensed along the range, while the wave spectra information is a function of SAR azimuth direction. The retrieving sea surface flows are delivered inshore zone of the coastal water of the Teluk Kemang, Port Dickson as part of the Malacca Straits.

Therefore, the Doppler shift frequency of the ATI indicates fluctuations of sea surface flow. The inshore water has a weak flow along 5 km of the coastal water. This is indicated by the lower rate value of 0.1 m/s. In this regard, the lowest spectral peak of the Doppler frequency shift is 0.04 which is corresponding to the frequency shift value of −200 Hz (**Figure 4**). In this view, the weak inshore water flow could be attributed to the impact of the low tide of 0.3 m as noticed along the coastal water of the Teluk Kemang, Port Dickson.

The interferogram phase is ranged between −0.7° and +0.7° (**Figure 5**) derived by the multichannel MAP height estimator. Obviously, the same pattern is visible.

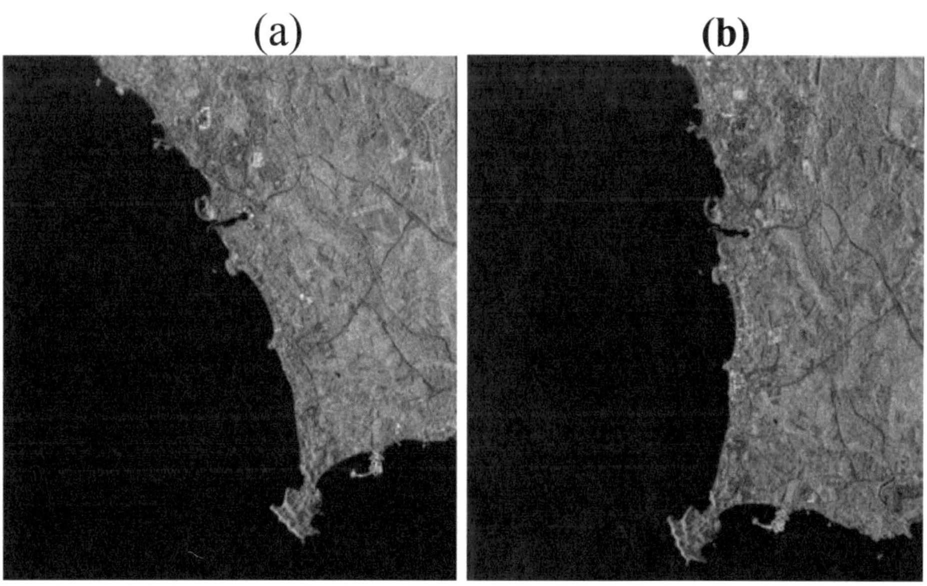

Figure 3.
TanDEM-X SAR data (a) first mission and (b) second mission with VV polarization.

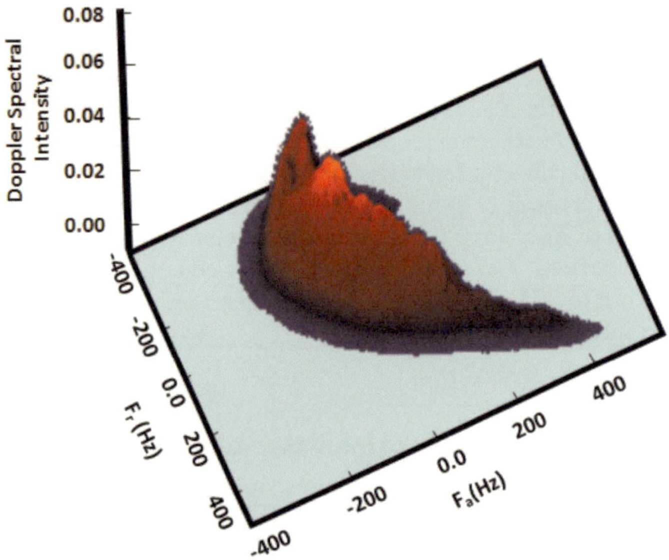

Figure 4.
TanDEM-X data Doppler spectra intensity.

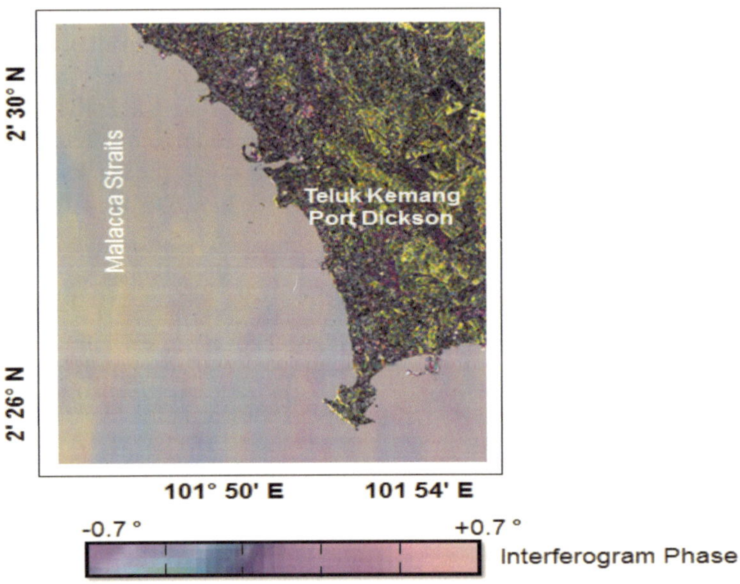

Figure 5.
Pattern of interferogram phase.

This pattern signature represents current feature variations along the coastal waters. Conversely, this interferogram phase is dominated by noises.

The inverters of the interferogram phase can be used to compute the ATI Doppler sea surface current. The ATI Doppler shows a clear current pattern movement along the coastal waters with minimum and maximum speed of 0.1 and 0.2 m/s, correspondingly (**Figure 6**). In fact, the interferometric combination of the two images reveals phase alterations that are comparable to the backscatter variations of the Doppler frequency shift [7]. This rapidity is conforming to sea-level differences of 0.4 m (**Figure 6**).

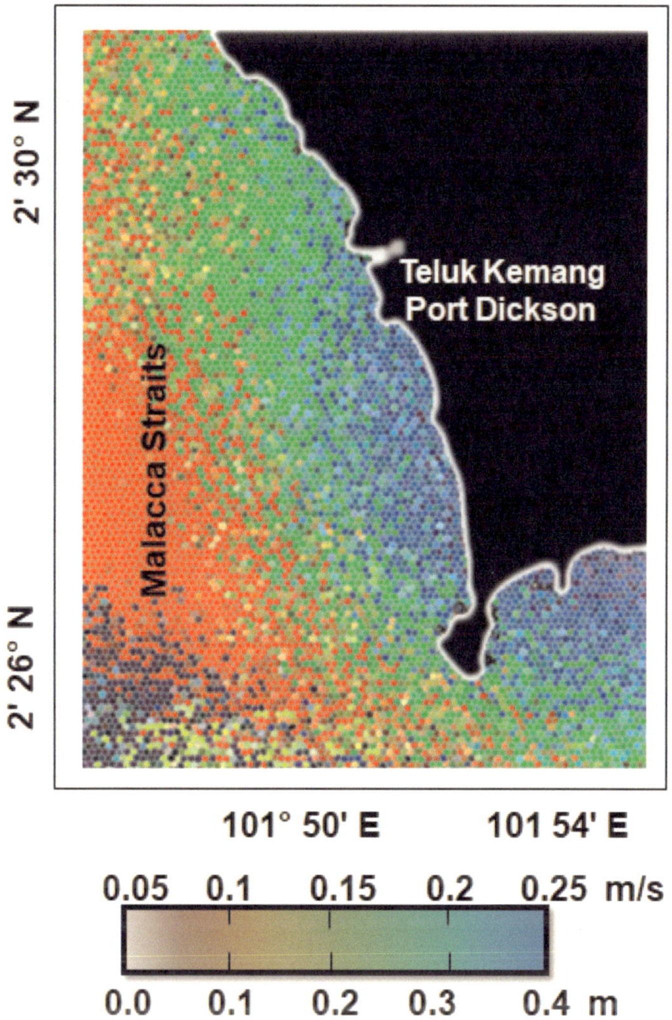

Figure 6.
Sea surface current and sea-level variations retrieved using an ATI MAP algorithm.

Figure 6 exhibits noteworthy correspondences between the consequence of sea surface flow speeds, which are created from TanDEM-X satellite data, and the consequence delivered in the in situ quantity. **Figure 7** demonstrates how the correlation coefficient alteration as direct correlation between the two different parameters is modified. Indeed, the investigation of the correlation between different measured parameters can assist to develop accurate model. Obviously, there is a worthy correlation between the retrieved sea surface flow and real in situ measured flow with r^2 of 0.76. Conversely, this correlation is not faultless, but it appears to have a confident, direct association, and resembles to what one would guess when bearing in mind both sea surface flow simulation from satellite data and one is measured in situ and then follow the hypothesis of normality.

As said by Romeiser et al. [6, 7], the signatures of the Doppler frequency shift are clearly responsive to sea surface flow than to wind modifications. Similarly, a modification of the Doppler frequency shifts has a tiny effect on the TanDEM-X backscatter intensity as compared to relaxation rate. Obtaining phase by using multichannel MAP height estimator algorithm will allow us to characterize the water sea-level fluctuations. Three-dimensional reconstruction of water-level

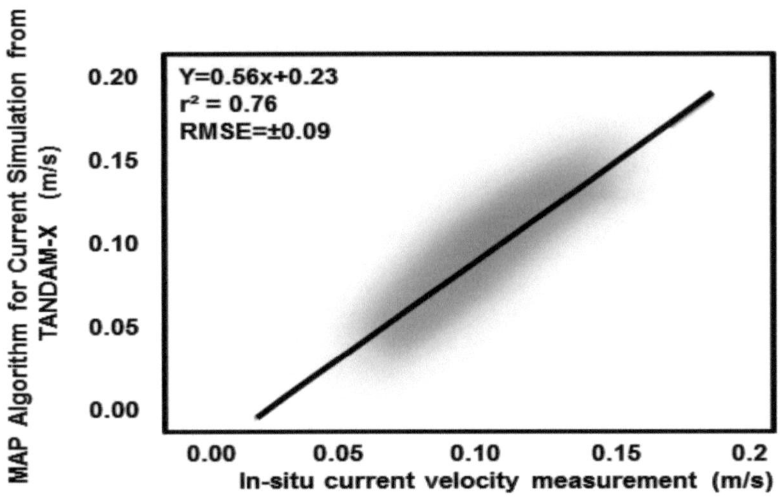

Figure 7.
Validation of MAP algorithm with in situ measurement.

changes from the ATInSAR technique by using the algorithm of multichannel MAP height approximation can aid to regulate the vertical shift of sea-level changes. Moreover, the multichannel MAP height approximation has achieved the difficulty of the phase unwrapping discontinuities and amended the vertical displacement synopsis as rivaled to conservative algorithm of phase unwrapping, for instance, (i) minimum-norm algorithm and (ii) path-following algorithm [9]. Lastly, TanDEM-X satellite data are comprehended as the prospective radar device for observing the dynamic fluctuation of ocean surface. Sea surface flow is considered as one of a consideration—impressing issue which is required a short visit cycle and extraordinary resolution. In this understanding, these can afford specific facts in relation to sea surface dynamic flow [2, 3, 5, 9–16].

5. Conclusion

This work has revealed a method for regaining sea surface flow using such high-resolution satellite data of TanDEM SAR-X. Along-track interferometry (ATI) technique is implemented to retrieve sea surface current movement. To this end, multichannel MAP height estimator algorithm is said to model sea-level variation. Then, the inverse algorithm is used which is based on the Doppler frequency model to retrieve sea surface current. The results reveal that the sea surface flow pattern is dominated by low velocity of less than 0.3 m/s which corresponds to lower sea-level variation of 0.4 m. The study confirms that multichannel MAP height estimator algorithm is proficient to regain the sea surface flow rate from ATI TanDEM-X with an extraordinary precision of ±0.09 m/s. In conclusion, the approximation algorithm of multichannel MAP height conceivably can be a tremendous practice for repossessing sea surface flow pattern and sea-level fluctuations from ATI TanDEM-X satellite data.

Author details

Maged Marghany
Faculty Geospatial and Real Estate, Geomatika University College, Kuala Lumpur, Malaysia

*Address all correspondence to: magedupm@hotmail.com

References

[1] Lu Z, Kim J-W, Lee H, Shum C, Duan J, Ibaraki M, et al. Helmand river hydrologic studies using ALOS PALSAR InSAR and ENVISAT altimetry. Marine Geodesy. 2009;**32**(3):320-333

[2] Ferraiuolo G, Meglio F, Pascazio V, Schirinzi G. DEM reconstruction accuracy in multichannel SAR interferometry. IEEE Transactions on Geoscience and Remote Sensing. 2009; **47**(1):191-201

[3] Marghany M. Developing robust model for retrieving sea surface current from RADARSAT-1 SAR satellite data. International Journal of Physical Sciences. 2011;**6**(29):6630-6637

[4] Mason DC, Speck R, Devereux B, Schumann GJ-P, Neal JC, Bates PD. Flood detection in urban areas using TerraSAR-X. IEEE Transactions on Geoscience and Remote Sensing. 2010; **48**(2):882-894

[5] Marghany M. Three-dimensional visualisation of coastal geomorphology using fuzzy B-spline of dinsar technique. International Journal of Physical Sciences. 2011;**6**(30):6967-6971

[6] Romeiser R, Runge H. Theoretical evaluation of several possible along-track InSAR modes of TerraSAR-X for ocean current measurements. IEEE Transactions on Geoscience and Remote Sensing. 2007;**45**(1):21-35

[7] Romeiser R, Runge H, Suchandt S, Kahle R, Rossi C, Bell PS. Quality assessment of surface current fields from TerraSAR-X and TanDEM-X along-track interferometry and Doppler centroid analysis. IEEE Transactions on Geoscience and Remote Sensing. 2014; **52**(5):2759-2772

[8] Yoon G-W, Kim S-W, Lee Y-W, Won J-S. Measurement of the water level in reservoirs from TerraSAR-X SAR interferometry and amplitude images. Remote Sensing Letters. 2013; **4**(5):446-454

[9] Baselice F, Ferraioli G, Pascazio V. DEM reconstruction in layover areas from SAR and auxiliary input data. IEEE Geoscience and Remote Sensing Letters. 2009;**6**(2):253-257

[10] Ferraiuolo G, Pascazio V, Schirinzi G. Maximum a posteriori estimation of height profiles in InSAR imaging. IEEE Geoscience and Remote Sensing Letters. 2004;**1**(2):66-70

[11] Marghany M. Three-dimensional coastal geomorphology deformation modelling using differential synthetic aperture interferometry. Zeitschrift fur Naturforschung A: Journal of Physical Sciences. 2012;**67**(6):419

[12] Marghany M. DEM reconstruction of coastal geomorphology from DINSAR. In: Murgante B et al., editors. Lecture Notes in Computer Science (ICCSA 2012). Part III, LNCS 7335. New York City: Springer; 2012. pp. 435-446

[13] Marghany M. DInSAR technique for three-dimensional coastal spit simulation from radarsat-1 fine mode data. Acta Geophysica. 2013;**61**(2):478-493

[14] Marghany M. Simulation of Tsunami Impact on Sea Surface Salinity along Banda Aceh Coastal Waters, Indonesia. In: Marghany M, editors. Advanced Geoscience Remote Sensing. Croatia: Intech; 2014. pp. 229-251

[15] Marghany M. Hybrid genetic algorithm of interferometric synthetic aperture radar for three-dimensional coastal deformation. In: Hybrid Genetic Algorithm of Interferometric Synthetic Aperture Radar for Three-Dimensional Coastal Deformation. 2014. pp. 116-131

[16] Marghany M. Simulation sea surface current from RADARSAT-2 SAR data using Hopfield neural network. In: Synthetic Aperture Radar (APSAR), 2015 IEEE 5th Asia-Pacific Conference on Synthetic Aperture Radar (APSAR). New York: Institute of Electrical and Electronics Engineers (IEEE); 2015. pp. 805-808

Chapter 2

On Feature-Based SAR Image Registration: Appropriate Feature and Retrieval Algorithm

Dong Li, Yunhua Zhang and Xiaojin Shi

Abstract

An investigation on the appropriate feature and parameter retrieval algorithm is conducted for feature-based registration of synthetic aperture radar (SAR) images. The commonly used features such as tie points, Harris corner, SIFT, and SURF are comprehensively evaluated. SURF is shown to outperform others on criteria such as the geometrical invariance of feature and descriptor, the extraction and matching speed, the localization accuracy, as well as the robustness to decorrelation and speckling. The processing result reveals that SURF has nice flexibility to SAR speckles for the potential relationship between Fast-Hessian detector and refined Lee filter. Moreover, the use of Fast-Hessian to oversampled images with unaltered sampling step helps to improve the registration accuracy to subpixel (i.e., <1 pixel). As for parameter retrieval, the widely used random sample consensus (RANSAC) is inappropriate because it may trap into local occlusion and result in uncertain estimation. An extended fast least trimmed squares (EF-LTS) is proposed, which behaves stable and averagely better than RANSAC. Fitting SURF features with EF-LTS is hence suggested for SAR image registration. The nice performance of this scheme is validated on both InSAR and MiniSAR image pairs.

Keywords: extended fast least trimmed squares (EF-LTS), feature-based image registration, parameter estimation, speeded up robust feature (SURF), synthetic aperture radar (SAR)

1. Introduction

Synthetic aperture radar (SAR) as an irreplaceable remote sensing technique has been used for earth observation and environment monitoring for a long time due to its all-weather and all-day operational capability. A large number of airborne and spaceborne SAR sensors have been deployed recently. Nevertheless, the difference in sensors and imaging geometries will always introduce a geometrical warp between images which should be compensated before any joint application of multiple SAR images for accurate apperception and understanding of target and scene. Image registration is just dedicated to retrieve the warp function to align the same pixel position in each SAR image to the same target in the global system.

A lot of SAR image registration techniques have been developed hitherto. In this chapter, we focus on the algorithms that conduct registration based on image features, such as contour, region, line, and point. Contour, region, and line as well as their combination are often used for registration of multi-modality images. For SAR images with geometrical distortion and speckle, point feature is generally much clearer and easier extracted. Tie points, corner, and keypoint are the commonly used features in SAR image registration. Tie points usually refer to the features extracted from tie patches in SAR image registration [1–4]. The tie patches are first matched by region-based algorithms, and the tie points are then located by extracting the geometrical centers or centroids of the matched patches. Corner denotes another kind of point feature which has two dominant but different edge directions in local neighborhood. In SAR image registration, Harris corner [5] is the commonly used point feature [2, 6] whose response function is the weighted addition of the determinant and squared trace of the first-order moment matrix which describes the local neighboring gradient distribution of a point. Keypoint refers to the point differing in brightness or color compared with the surrounding. It is identified to further enable a complementary description of image structure that cannot be characterized by corner. The scale invariant feature transform (SIFT) [7] and the speeded up robust feature (SURF) [8] are the widely used keypoints in SAR image registration. SIFT was developed by Lowe [7] to extract features based on the automatic scale selection theory. Lindeberg [9] found that the only possible scale-space kernel under a variety of reasonable assumptions is the Gaussian function, and he experimented with both the traces of Hessian matrix, i.e., the Laplacian of Gaussian (LoG) and the determinant of Hessian (DoH) matrix, to detect the blob-like structures. To extract keypoints efficiently, Lowe [7] simplified LoG with the difference of Gaussian (DoG) further. SIFT enables not only a feature detector, but also a 128D vectorized descriptor of gradient and orientation. Mikolajczyk and Schmid conducted a comparative study on 10 different local descriptors and found that SIFT performs the best on treating the common image deformations [10]. SIFT has been widely used in SAR image registration [11–23]. Chen et al. [13] systematically evaluated the application of SIFT to SAR and displayed its usefulness for image registration. Schwind et al. [15] further indicated that SIFT is a robust alternative for point feature-based SAR image registration. The bottleneck of SIFT is the speed [8, 13, 15], which hinders its application to general SAR image registration. To accelerate SIFT, Schwind et al. [15] proposed to skip features detected at the first octave of the scale space pyramid (SSP) because matches extracted from this octave have the highest matching false alarm rate (MFAR). This can save the processing time without reducing the number of correct matches greatly. However, the first scale octave in SSP of SIFT refers to the image of original size or doubled size which has the highest resolution in SSP. Thus, the features extracted from this octave are more accurate for image registration [16]. Therefore, the discarding of matches from the first octave may influence the final registration accuracy. Based on the same scheme as SIFT, SURF developed by Bay et al. [8] uses a combination of novel detection, description, and matching methods to simplify SIFT. SURF extracts feature based on DoH instead of its trace because DoH bears slightly better scale selection property under non-Euclidean affine transformation than LoG. Bay et al. used a Fast-Hessian detector with box filters to approximate DoH. The SURF descriptor is a 64D vector composed by the Harr wavelet responses of the square area around keypoint. SURF has been demonstrated to outperform SIFT on speed, repeatability, distinctiveness, and robustness [8]. It has been used for multispectral satellite image registration [24], seabed recognition based on sonar images [25], and SAR image registration [26–29].

The next procedure after feature extraction is to match the features for correspondences. For tie points, this procedure is unnecessary because they have already matched when extracted. For other features, the correspondences are usually constructed by optimizing certain merit function, such as maximizing the similarity or minimizing the difference. The warp function can then be retrieved by fitting the obtained correspondences. For correspondences without any mismatches, the retrieval can be easily conducted by fitting them with the least squares (LS). However, for the general registration cases, the initial correspondences often contain mismatches. Therefore, the robust retrieval algorithms which are insensitive to outliers are needed. In many existing literatures on feature-based SAR image registration [15, 16, 26, 27], the random sample consensus (RANSAC) [30] has been widely used and recommended for warp function retrieval. RANSAC conducts the estimation by randomly sampling a minimal sampling set (MSS) to achieve an estimation of the warping, and the entire datasets are then checked on the estimation for a consensus set (CS) of correspondences. These two steps are iterated until the largest CS is achieved [31]. Besides this, the least median squares (LMedS) [32] and the fast least trimmed squares (Fast-LTS) [33] have also been used [4, 34, 35]. There are also some other approaches which use different matching and retrieval algorithms with different features, which can be referred to the related reviewing articles [36–38].

Although lots of approaches have been developed for feature-based SAR image registration, there are still some open problems that have not been perfectly solved yet. In this chapter, we concentrate on two problems, i.e., which feature is more appropriate and which retrieval algorithm performs much better? The first problem is related to the feature operator, which is focused in Sections 2 and 3. We give a detailed evaluation to tie points, Harris corner, SIFT, and SURF in terms of the geometrical invariance of feature and descriptor, extraction and matching speed, localization accuracy, robustness to decorrelation, and flexibility to speckle. SURF is identified to outperform others. Particularly, we find that SURF is flexible to speckle for the close relationship between Fast-Hessian detector and refined Lee speckle filter. SURF is thus more competent for SAR image registration. The second problem is posed in Section 4 with the reason that the widely used RANSAC is found instable for parameter estimation in the registration of an interferometric SAR (InSAR) image pair. The uncertainty arises from its inappropriate loss function and estimation strategy. Based on the scheme of Fast-LTS, an extended Fast-LTS (EF-LTS) is presented for 2D robust parameter estimation. Experiment on InSAR image pair demonstrates that EF-LTS is more stable and robust than RANSAC. It is more appropriate and competent for SAR image registration. Based on these, we recommend fitting the SURF features with EF-LTS to conduct the registration. We further evaluate this scheme in Section 5 by processing the MiniSAR image pair, and the result complies with our expectation. Section 6 concludes the chapter finally.

2. Comparative analysis on the commonly used features for SAR image registration

SAR image is acquired with intensity and phase, which should be transformed into the real one before feature detection by taking the intensity or the logarithmic intensity of the image. Instead of proposing a novel feature for SAR image registration, we identify the appropriate feature from the widely used tie points, Harris corner, SIFT, and SURF by evaluating them on several criteria. In this section, the features will be evaluated on the following six factors, i.e., the geometrical invariance of feature, the extraction speed, the localization accuracy, the geometric invariance of descriptor, the matching speed, and the robustness to decorrelation, while the impact of SAR speckles will be particularly focused and analyzed in Section 3.

2.1 Geometrical invariance of feature

The geometrical invariance of feature refers to which degree of warping a same feature can still be extracted from the warped images by a detector. Cross-correlation (CC) is sensitive to image rotation and scaling, hence the CC-based tie points are only invariant to the following translation transformation:

$$\begin{bmatrix} x' \\ y' \\ 1 \end{bmatrix} = \begin{bmatrix} 1 & 0 & t_x \\ 0 & 1 & t_y \\ 0 & 0 & 1 \end{bmatrix} \begin{bmatrix} x \\ y \\ 1 \end{bmatrix} \tag{1}$$

where $(x, y, 1)^{\mathrm{T}} \leftrightarrow (x', y', 1)^{\mathrm{T}}$ are the inhomogeneous coordinates of a pair of matching points (the superscript T shows the vector transpose), and t_x and t_y denote the translations in x- and y-direction, respectively. The Harris measure is the following Harris matrix **H** describing the neighboring gradient distribution of a point [5]:

$$\mathbf{H} = \begin{bmatrix} \langle I_x^2 \rangle & \langle I_x I_y \rangle \\ \langle I_x I_y \rangle & \langle I_y^2 \rangle \end{bmatrix} \tag{2}$$

where $<\cdot>$ denotes the ensemble average; I_x and I_y are the first-order partial derivatives in x- and y-direction, respectively. Then, the response function R of Harris is the weighted sum of the determinant and squared trace of **H** [5]:

$$R = \det(\mathbf{H}) - \kappa(\mathrm{trace}(\mathbf{H}))^2 \tag{3}$$

where the weight κ is a constant within the interval 0.04–0.06. A pixel is selected as a Harris corner if its response R is beyond a given threshold. It can be easily obtained from (2) that **H** is semi-definite Hermitian, which indicates the existence of two nonnegative eigenvalues λ_1 and λ_2. Then (3) can be further formulated as:

$$R = \lambda_1\lambda_2 - \kappa(\lambda_1 + \lambda_2)^2 \tag{4}$$

The Harris response R is only decided by the eigenvalues of **H**. Any unitary transformation of **H** will not influence the extraction of corner. Therefore, Harris corner is invariant to the following Euclidean transformation:

$$\begin{bmatrix} x' \\ y' \\ 1 \end{bmatrix} = \begin{bmatrix} \cos\theta & -\sin\theta & t_x \\ \sin\theta & \cos\theta & t_y \\ 0 & 0 & 1 \end{bmatrix} \begin{bmatrix} x \\ y \\ 1 \end{bmatrix} \tag{5}$$

where θ denotes the rotation. SIFT and SURF were proposed to achieve the scale-invariance further:

$$\begin{bmatrix} x' \\ y' \\ 1 \end{bmatrix} = \begin{bmatrix} s\cos\theta & -s\sin\theta & t_x \\ s\sin\theta & s\cos\theta & t_y \\ 0 & 0 & 1 \end{bmatrix} \begin{bmatrix} x \\ y \\ 1 \end{bmatrix} \tag{6}$$

where s is the scale. Theoretically, SIFT and SURF features are not affine-invariant as Harris-Affine and Hessian-Affine features [39]. Nonetheless, the affine frame in Hessian-Affine and Harris-Affine is more sensitive to noise than scale-invariant detector. For general SAR image application, scale-invariant features such as SIFT and SURF are sufficient.

2.2 Feature extraction speed

The extraction speed is mainly influenced by the computational load of detector. Tie points are identified by traversing all potential offsets to calculate CC. The resulted computational load is heavy. The Harris response R is determined by the determinant and trace of matrix $\mathbf{H}$. The calculation of $\mathbf{H}$ only relates to the first-order derivatives which can be fast achieved. The scale-invariant SIFT and SURF keypoints are extracted by constructing SSP first. SSP is comprised of several octaves and each octave consists of several scale levels further. A scale level is a Gaussian-smoothed image. The nearby two layers are subtracted to calculate DoG, an approximation to LoG. The keypoint is finally identified as the point with extreme value of DoG in a $3 \times 3 \times 3$ neighborhood in the scale space. SIFT detector performs slower than Harris because it extracts the feature in 3D space not in 2D space. Nonetheless, to extract the same number of subpixel features, SIFT detector is faster than CC-based tie points for the latter conducts exhaustive searching. SURF extracts feature based on DoH. Given a point $\mathbf{x} = (x, y)$ in image $\mathbf{I}$ at scale σ, the scale function DoH is obtained by:

$$DoH_{SIFT} = L_{xx}(\mathbf{x}, \sigma)L_{yy}(\mathbf{x}, \sigma) - \left(L_{xy}(\mathbf{x}, \sigma)\right)^2 \tag{7}$$

where L_{xx} $(\mathbf{x}, \sigma)$, L_{yy} $(\mathbf{x}, \sigma)$, and L_{xy} $(\mathbf{x}, \sigma)$ denote the convolution of the Gaussian second-order derivative in x-, y-, and xy-directions with $\mathbf{I}$, respectively.

When applied in practice, Gaussians should be discretized and cropped. The corresponding discretized and cropped L_{xx}, L_{xy}, and L_{yy} with the lowest scale of 1.2 are displayed in the first row of **Figure 1**. Encouraged by the successful simplification of LoG with DoG in SIFT, Bay et al. devised a Fast-Hessian detector to approximate L_{xx}, L_{xy}, and L_{yy} with box filters D_{xx}, D_{xy}, and D_{yy}, respectively, shown in the second row of **Figure 1**. In [8], Bay et al. indicated that the performance of this approximation is comparable or even better than the original Gaussians. The

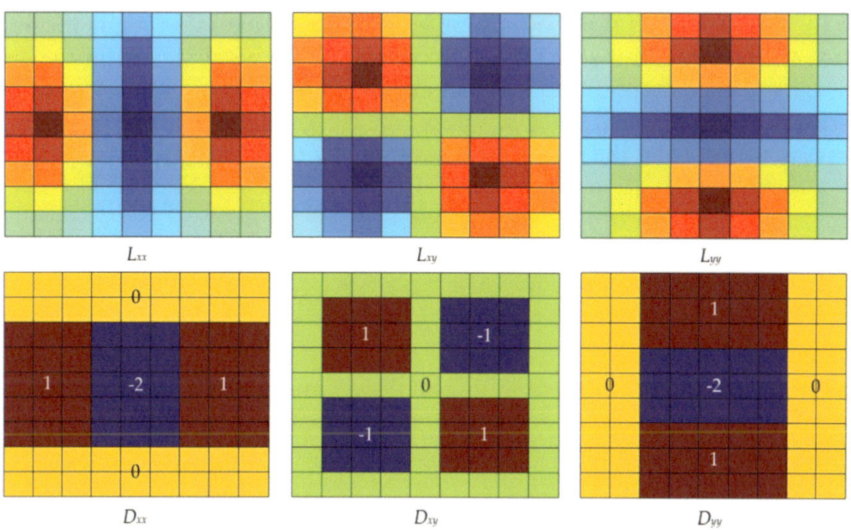

Figure 1.
SIFT discretized and cropped Gaussian second-order partial derivatives in x- (L_{xx}), xy- (L_{xy}), and y-direction (L_{yy}), as well as their corresponding SURF box filter approximations D_{xx}, D_{xy}, and D_{yy}, respectively.

approximation makes pixels in certain window have the same weight. The convolutions can be then calculated at very low computational cost by using the integral image. Therefore, instead of iteratively reducing the image size and using the cascade filtering, SSP in SURF is built by simply up-scaling the box filters without changing the size of the image. The use of integral image enables the convolutions independent of the filter size and scale.

2.3 Localization accuracy of feature

Image registration accuracy is closely determined by the localization accuracy of feature. Tie points achieve subpixel accuracy by oversampling the image patches [40] or CC obtained in coarse registration [41]. Higher sampling rate indicates higher accuracy, but it also signifies larger data sets, heavier computational load, and more severe aliasing. Keypoint in SIFT and SURF is first located as the extrema using the non-maximum suppression technique, and is then refined to subpixel and sub-scale accuracy by Taylor fitting a 3D quadratic to the scale function DoG (for SIFT) or the approximated DoH (for SURF) in the scale space [42]:

$$f(\mathbf{X}) = f(\mathbf{X}_0) + \left(\frac{\partial f}{\partial \mathbf{X}}(\mathbf{X}_0)\right)^{\mathrm{T}} \Delta\mathbf{X} + \frac{1}{2}\Delta\mathbf{X}^{\mathrm{T}}\left(\frac{\partial^2 f}{\partial \mathbf{x}^2}(\mathbf{X}_0)\right)\Delta\mathbf{X}. \tag{8}$$

Therefore, SIFT and SURF can obtain the highest accuracy. However, it should be noted that although the subpixel feature localization is the precondition of accurate image registration, it cannot guarantee a subpixel image registration. For high accurate SAR image registration, we should further evaluate the features carefully, and this will be detailed in Section 3.4.

2.4 Geometrical invariance of descriptor

Feature descriptor is usually a vector depicting the neighboring information of a feature. It plays a key role in feature matching. The descriptor's geometrical invariance determines the degree of warping to which features can still be successfully matched. Harris corner and tie points have no descriptor. From feature matching point of view, however, they both adopt template matching by selecting the image square centered around the feature as descriptor, which is only invariant to translation. Thus, tie points and Harris corner can be successfully matched only under weak warping. SIFT and SURF descriptors enable a good compromise between feature complexity and the robustness to commonly occurring deformation such as weak affine transformation [7, 8, 43]:

$$\begin{bmatrix} x' \\ y' \\ 1 \end{bmatrix} = \begin{bmatrix} s_x \cos\theta & -s_y \sin\theta & t_x \\ s_x \sin\theta & s_y \cos\theta & t_y \\ 0 & 0 & 1 \end{bmatrix} \begin{bmatrix} x \\ y \\ 1 \end{bmatrix} \tag{9}$$

where s_x and s_y denote the scales in directions x and y, respectively. Robust matching across a substantial range of affine distortion and change in 3D viewpoint can hence be achieved.

2.5 Matching speed of feature

Feature matching is usually conducted based on certain merit function of the descriptors. In feature-based SAR image registration, the merit function is to maximize the similarity (such as CC [4]) or minimize the differences (such as Euclidean distance [7, 8]). A correspondence is detected if it can optimize the merit function. For SIFT and SURF, the merit of an optimal correspondence has also to be certain times larger than the second optimal merit. Matching speed is mainly determined by the calculation of merit. For tie points and Harris corner, the merit function is the maximum of CC, which can be obtained on complex data or magnitude data [44], referring to coherent CC or incoherent CC, respectively. The registration accuracy attained by coherent CC is much higher than that by incoherent CC [45]. If D_1 and D_2 are the image patches, respectively, centered at an initial match, the coherent CC is calculated as

$$CC(D_1, D_2) = \frac{\left|\sum_{i=1}^{N}\sum_{j=1}^{N}(D_1(i,j) - \mu_1)(D_2(i,j) - \mu_2)^*\right|}{\sqrt{\sum_{i=1}^{N}\sum_{j=1}^{N}|(D_1(i,j) - \mu_1)|^2\sum_{i=1}^{N}\sum_{j=1}^{N}|(D_2(i,j) - \mu_2)|^2}} \tag{10}$$

where N is the size of the image patch, μ_1 and μ_1 denote the means of D_1 and D_2, respectively. Equation (10) requires about $10N^2$ operations including $7N^2$ additions and $3N^2$ multiplications.

The merit function in SIFT and SURF is the minimum of the Euclidean distance. If D_3 and D_4 are the descriptors of an initial match, respectively, the distance can be calculated by

$$Dist(D_3, D_4) = \sum_{i=1}^{L} |D_3(i) - D_4(i)|^2 \tag{11}$$

where L is the length of descriptor. Equation (11) requires $3L$ operations including $2L$ additions and L multiplications. For SURF, Bay et al. [8] found that the sign of Laplacian can be further used to distinguish the feature from its background for fast indexing during matching stage. The merit will not be computed unless the initial match has the same sign. Hence, under the assumption of equal probability distribution for sign of Laplacian, the merit computation in SURF requires $1.5L$ operations. Taking the descriptor lengths L for SIFT and SURF being 128 and 64 into consideration, then (11) involves in 384 and 96 operations for SIFT and SURF, respectively. Hence, SURF is four times faster than SIFT on feature matching. To achieve the same efficiency as SIFT or SURF, the equivalent patch size N for tie points and Harris corner should be about 6 or 3, respectively. This may lead to biased CC estimation thus bad feature localization and matching due to the insufficient sampling.

2.6 Robustness to decorrelation

SAR decorrelation sources can be classified into two categories, i.e., the geometrical warping and radiometric warping. Geometrical warping will lead to decorrelation and influence the CC-based feature matching, which relates to the geometrical invariance of feature discussed above. Here, we focus on the radiometric warping-induced decorrelation. Such decorrelation is resulted because CC is only invariant to affine changes in scattering. Target scattering in microwave band is sensitive to frequency, bandwidth, and polarization. All these introduce a complex nonlinear radiometric warping, which degrades SAR information and aggravates image registration by impacting the localization of tie points. The localization accuracy of tie points is measured by the error standard deviation σ_L [45]:

$$\sigma_L = \sqrt{\frac{3}{2N^2}\frac{\sqrt{1-\gamma^2}}{\pi\gamma}osr^{3/2}} \tag{12}$$

where γ is CC, N is the size of the tie patches, and osr is the oversampling rate. Localization accuracy directly relates to CC: higher coherence means higher localization accuracy, while higher decorrelation indicates worse localization accuracy and worse registration accuracy. It is known that one can approximate a nonlinear function with a series of linear functions, so a nice method to improve the robustness to decorrelation is to use smaller image patches, but this will also result in worse localization accuracy through N in (12). Thus, tie points are not robust to decorrelation. Similarly, the influence of decorrelation on CC-based matching of Harris corners is also unavoidable. However, Harris, SIFT, and SURF locate feature based on geometrical texture instead of correlation. This will reduce the influence of decorrelation. The matching of SIFT and SURF features is based on local descriptors which are invariant to affine changes in scattering. SIFT and SURF features are thus more robust to decorrelation.

3. Impact of SAR speckles on accurate feature extraction

SAR image is acquired by actively measuring and coherently processing the electromagnetic scattering of target. The interference of scatterings from scatterers within each resolution cell produces a pixel-to-pixel variation in image intensity and results in the so-called speckle. In this section, we first conduct a qualitative evaluation on the flexibility of existing features to speckles. An experimental evaluation of the identified feature is then conducted and some necessary improvements are developed for high accurate SAR image registration.

3.1 Flexibility to image speckling

For CC-based tie points, the assumption that the scattering is locally stationary and ergodic may not be tenable in the existence of speckles. As a result, the correlation estimation as well as the localization and matching of the feature will be biased. For the geometrical texture-based detectors such as Harris, SIFT, and SURF, speckles may lead to false texture and high MFAR. To achieve stable features from the speckle-contaminated SAR image, a conceivable method is to suppress speckle beforehand. Schwind et al. [15] suggested adopting the ISEF filter, but they indicated that ISEF filter and any other filter may slightly affect feature localization and registration quality. Hence, a better strategy is to conduct speckle suppression while feature extraction, i.e., the detector should be flexible to speckling.

Harris detector obtains features using the first-order image derivatives which are not robust to speckles. As a result, Harris detector may extract many features, but most of the extracted features are speckles with only a few correct matches. This influence has been also observed by Schwind et al. [15] when using SIFT to SAR: only very few matches are constructed at the first octave of SSP although with extensive number of extractable features, and the matches from this octave have the highest MFAR of all the octaves. The first scale octave refers to the original or double-sized images which are of the highest resolution and the largest number of extractable keypoints. The highest MFAR at this octave clearly indicates the bad flexibility of SIFT to speckles, while the lower MFAR at higher octaves is just due to the fact that larger image smoothing reduces the speckle. Different from SIFT, SURF can deal with speckle very well because of the relationship between Fast-Hessian detector and refined Lee speckle filter.

3.2 Refined Lee speckle filter

An ideal speckle filter should adaptively smooth speckle, retain the sharpness of boundaries and edges, and preserve the subtle but distinguishable details. The most widely used boxcar filter replaces a pixel with the mean of its windowed neighborhood. This filter can be easily implemented and works very well in homogeneous area, but will degrade spatial resolution in inhomogeneous area due to the indiscriminate averaging [46]. To solve this, many filtering techniques have been proposed. The refined Lee speckle filter is just such a filter which uses the local statistics to suppress speckles without degrading image. To identify pixels with the similar texture, Lee devised the eight non-square edge-aligned windows, as shown in **Figure 2**. In the course of filtering, one of the windows is matched to calculate local statistics based on edge direction, and the minimum mean square algorithm is then adopted for filtering. As a result, this filter can effectively reduce the speckle without degrading the edge [46].

3.3 Relationship between Fast-Hessian detector and refined Lee filter

As mentioned previously, SURF extracts features based on the box filter displayed in **Figure 1**. Box filter not only speeds up feature extraction, but also enables SURF to extract features while reducing speckles. In D_{xx} of **Figure 1**, we average the pixels using a 5×3 window first, and then extract the vertical edge by the second-order image partial derivative in x-direction with convolution template $[1\ {-2}\ 1]$. This is equivalent to filter speckles with Lee's windows **Figure 2(a)** and **(e)**. Similarly, D_{yy} denotes that we also filter the pixels using a 5×3 window first, but then extract the horizontal edge using the second-order image partial derivative in y-direction with convolution template $[1\ {-2}\ 1]^{T}$. This is equivalent to filter speckle with Lee's non-square windows **Figure 2(c)** and **(g)**. D_{xy} shows that we use a 3×3 window and extract the 135° edge feature by the second-order image partial derivative in negative xy-direction with the convolution template $[1\ {-1};\ {-1},\ 1]$. This is equivalent to filter speckle with windows **Figure 2(d)** and **(h)**. Likewise, $-D_{xy}$ gives that we also use a 3×3 window but extract the 45° edge by the second order image partial derivative in positive xy-direction with convolution template $[-1\ 1;\ 1,\ {-1}]$. This is equivalent to filter with windows **Figure 2(b)** and **(f)**. Instead of

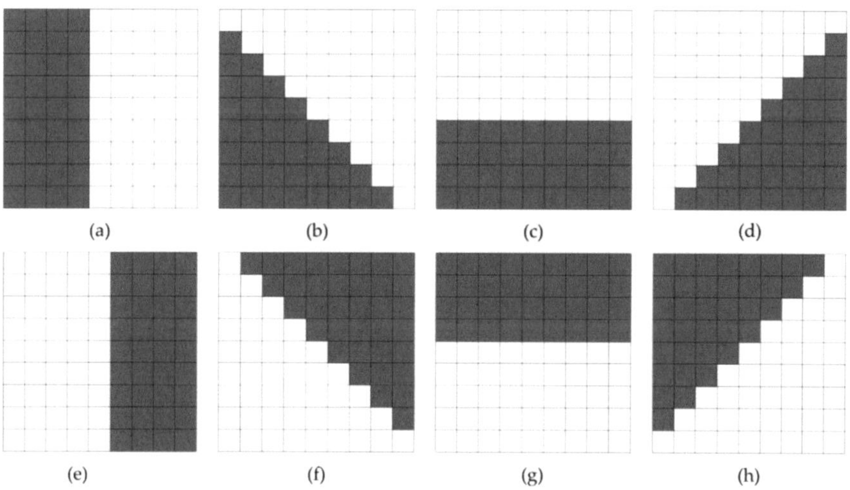

Figure 2.
Edge-aligned windows used in refined Lee filter to decide the local texture, where windows (a) and (e) are used for vertical edge, (c) and (g) for horizontal edge, (b) and (f) for 135° edge, and (d) and (h) for 45° edge. The pixels in white are used for filtering computation.

selecting the optimal edge to calculate local statistics, the four edge features are combined to a new feature in SURF by:

$$DoH_{SURF} = D_{xx}D_{yy} + (0.9D_{xy})(-0.9D_{xy}) \tag{13}$$

which corresponds to DoH in (7), where the constant 0.9 is used to balance the expression for the Hessian's determinant. Then, SSP in SURF just indicates that we adopt a series of box filters of different size to filter speckles and extract features of different scales. Hence, SURF is very flexible to deal with speckle.

3.4 Evaluation of SURF for SAR image subpixel registration

As listed in **Table 1**, according to the comparative analysis in Sections 2 and 3.1 on several criteria, we can obtain that for the general registration of SAR images

- SURF outperforms others in terms of the considered criteria.
- SIFT is applicable when no strict requirement for speed.
- Harris may be appropriate for coarse registration.
- Tie points are fit for images with slight distortion and weak decorrelation and require heavy computation load.

From these, we can see that SURF is more appropriate and competent for general SAR image registration. Nevertheless, SAR applications, like DEM retrieval and deformation estimation usually impose a strict requirement for registration accuracy. To ensure an acceptable result, the registration accuracy should be subpixel. To evaluate the capability of SURF for subpixel image registration, we devise a comparative experiment on some contrived SAR image pairs. **Figure 3** shows a SAR image of Enta Volcano acquired by SIR-C/X-SAR. We treat this image as the master and transform it to model an affine geometrical warp for the slave image.

Items	Tie points	Harris corner	SIFT	SURF
Geometrical invariance of feature	Translation	Rotation and translation	Scaling, rotation, and translation	Scaling, rotation, and translation
Feature extraction speed	Slower	Faster	Slow	Fast
Feature localization accuracy	Subpixel*	Pixel	Subpixel	Subpixel
Geometrical invariance of feature descriptor	Translation	Translation	Affine transform	Affine transform
Feature matching speed	Slow	Slow	Fast	Faster
Robustness to decorrelation	Worse	Bad	Good	Good
Flexibility to image speckle	Good	Bad	Bad	Better

**Determined by the sampling rate.*

Table 1.
Evaluation of the four commonly used features for SAR image registration in terms of several criteria.

$$\begin{bmatrix} x_s \\ y_s \\ 1 \end{bmatrix} = \mathbf{A} \begin{bmatrix} x_m \\ y_m \\ 1 \end{bmatrix} = \begin{bmatrix} a & b & t_x \\ c & d & t_y \\ 0 & 0 & 1 \end{bmatrix} \begin{bmatrix} x_m \\ y_m \\ 1 \end{bmatrix} \tag{14}$$

where $(x, y, 1)^{\mathrm{T}}$ are the homogenous image coordinates, subscripts s and m denote the slave and master images, respectively. **A** is an affine matrix composed by parameters a, b, c, and d, as well as two translations t_x and t_y. Bay et al. devised two versions of Fast-Hessian detectors for SURF. The one initializes SSP by using 9×9 box filter to the original image is denoted as *FH*-9(-1), while the one initializes SSP by using 15×15 box filter to double-sized image (also with doubled sampling step) is denoted as *FH*-15(-2). *FH*-15(-2) has been shown to be better than *FH*-9(-1) on repeatability [8]. We use the two detectors to extract point correspondences, respectively, based on which the robust EF-LTS (will be presented in Section 4) is then used to retrieve the warp matrix. To compare the two SURF detectors for SAR image registration, we consider four criteria, i.e., the average transfer error (*ATE*), *MFAR*, the number of correct matches, and the warp matrix estimation error (*WMEE*). *ATE* measures the appropriateness of the extracted features to the achieved warp parameters:

$$ATE = \frac{1}{N} \sum_{i=1}^{N} \left\| \begin{bmatrix} x_{si} \\ y_{si} \\ 1 \end{bmatrix} - \mathbf{A_r} \begin{bmatrix} x_{mi} \\ y_{mi} \\ 1 \end{bmatrix} \right\| \tag{15}$$

where $\mathbf{A_r}$ indicates the warp matrix retrieved on all the constructed correspondences (x_{si}, y_{si}) and (x_{mi}, y_{mi}) denote the ith correct correspondence located in slave image and master image, respectively, and N is the number of correct matches which are selected by:

$$\left\| \begin{bmatrix} x_{si} \\ y_{si} \\ 1 \end{bmatrix} - \mathbf{A} \begin{bmatrix} x_{mi} \\ y_{mi} \\ 1 \end{bmatrix} \right\| < threshold \begin{cases} \text{True} & (x_{si}, y_{si}) \leftrightarrow (x_{mi}, y_{mi}) \text{ is a correct match} \\ \text{False} & (x_{si}, y_{si}) \leftrightarrow (x_{mi}, y_{mi}) \text{ is a mismatch} \end{cases} \tag{16}$$

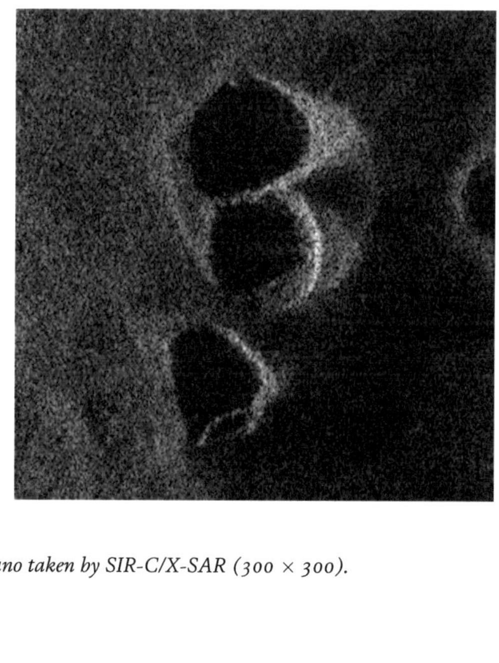

Figure 3.
SAR image of Enta Volcano taken by SIR-C/X-SAR (300 × 300).

where **A** is the true warp matrix. The *threshold* is chosen as 5 pixels, i.e., a correspondence is identified as a mismatch if the transfer error is larger than 5 pixels in any image direction.

MFAR, also called 1-precision [10], is defined as:

$$MFAR = \frac{\#matches - \#correct\ matches}{\#matches} \tag{17}$$

where "#" denotes "the number of." *MFAR* is just the rate of mismatches, which is related to image speckling as well as the radiometric and geometrical warping. It can be used together with *#correct matches* to evaluate the robustness of a detector to speckles on SAR image pair with controlled radiometric and geometrical warping.

WMEE is used to evaluate the consistency of the retrieved warp matrix and its true value:

$$WMEE = \|\mathbf{A} - \mathbf{A_r}\|_F \tag{18}$$

where $\|\cdot\|_F$ denotes the Frobenius norm.

We evaluate the two SURF detectors on four image pairs with different transformations, the retrieved warp matrix parameters, *ATE*, correct match number, *MFAR*, and *WMEE* are listed in **Table 2**. It shows that *FH*-15(-2) can extract more correct matches with lower *MFAR* than *FH*-9(-1). This validates the robustness of SURF to speckling because *FH*-15(-2) performs the feature extraction on the double-sized image with much serious speckle. *ATE* of *FH*-15(-2) is smaller than that of *FH*-9(-1) except on the first image pair. On all the four pairs, the features extracted by *FH*-15(-2) can obtain subpixel estimation in both image directions, but *FH*-9(-1) obtains this only on the first pair. Therefore, *FH*-15(-2) features are more consistent with the retrieval parameters. This also signifies that *FH*-15(-2) can attains lower *MFAR* than *FH*-9(-1) because parameter estimation in EF-LTS is related to the outlier percentage in data. This will be detailed in Section 4. As on *WMEE*, the two detectors perform equally, *FH*-15(-2) does not improve the registration accuracy on all the four pairs as we expected, and there is still clear inconsistency between the retrieved warp matrix and the true value. The reason lies in that the sampling step is also doubled when *FH*-15(-2) doubles the image. This makes sampling being still conducted on the equivalently same pixel position rather than the subpixel image position. For instance, let (x_0, y_0) be a sampled pixel in the original image, the corresponding position in doubled image is $(2x_0, 2y_0)$. The doubled step then makes this pixel position be still sampled instead of $(2x_0 \pm 1, 2y_0 \pm 1)$, while the latter corresponds to the subpixel position $(x_0 \pm 0.5, y_0 \pm 0.5)$ in the original image and positively contributes to the subpixel registration. Based on this, we suggest initializing SSP by using 9×9 box filter to the oversampled image but with unchanged sampling, we denote this detector as *FH*-9(-*Fs*), *Fs* denotes the sampling rate. To avoid nonlinear aliasing, the linear interpolator such as bilinear interpolator is used to conduct the sampling. **Table 2** further summarizes the registration results based on *FH*-9(-2) to *FH*-9(-5) detector. Comparing with *FH*-9 (-1) and *FH*-15(-2), the correct match number, *ATE*, *MFAR*, and *WMEE* of *FH*-9 (-2) are all clearly improved. As oversampling rate increases from 2 to 5, the registration accuracy is also improved for more correspondences of higher localization accuracy are identified. All these make the high accurate SAR image registration possible. In view of the fact that oversampling will increase dataset and computational load, for high accuracy registration we recommend oversampling the image three or four times so as to achieve the compromise among accuracy, robustness, and computational complexity.

Detectors	Estimated affine warping parameters						Correct match number and *MFAR*	*ATE*	*WMEE*
	a	*b*	*c*	*d*	t_x	t_y			
True value	0.7189	0.0452	−0.0402	0.8087	1.7000	2.4000	—	—	—
FH-15(-2)	0.7164	0.0415	−0.0481	0.8059	2.3151	3.6269	42 (0.1923)	(0.7109, 0.8770)	1.3725
FH-9(-1)	0.7195	0.0425	−0.0347	0.8067	2.0444	1.3480	22 (0.2414)	(0.5887, 0.7854)	1.1070
FH-9(-2)	0.7186	0.0458	−0.0395	0.8085	1.3565	2.2052	73 (0.1300)	(0.6219, 0.6602)	0.3949
FH-9(-3)	0.7192	0.0450	−0.0403	0.8088	1.4752	2.3425	129 (0.1164)	(0.3001, 0.4602)	0.2321
FH-9(-4)	0.7181	0.0453	−0.0402	0.8093	1.6070	2.1746	188 (0.1754)	(0.2580, 0.3790)	0.2439
FH-9(-5)	0.7186	0.0457	−0.0398	0.8094	1.4895	2.1085	176 (0.1619)	(0.2819, 0.3874)	0.3596
True value	0.9361	0.1889	−0.1617	1.0938	−10.5000	−3.4000	—	—	—
FH-15(-2)	0.9370	0.1887	−0.1576	1.0908	−10.5603	−3.7546	55 (0.0678)	(0.6040, 0.7075)	0.3598
FH-9(-1)	0.9298	0.1909	−0.1603	1.0868	−9.9304	−2.2059	25 (0.2188)	(0.7949, 1.2405)	1.3231
FH-9(-2)	0.9352	0.1898	−0.1618	1.0940	−10.4452	−3.4432	170 (0.0449)	(0.3821, 0.5200)	0.0698
FH-9(-3)	0.9361	0.1890	−0.1613	1.0937	−10.4329	−3.4817	419 (0.0141)	(0.2267, 0.3080)	0.1058
FH-9(-4)	0.9361	0.1890	−0.1617	1.0937	−10.4252	−3.4490	735 (0.0252)	(0.1703, 0.2143)	0.0894
FH-9(-5)	0.9360	0.1890	−0.1616	1.0938	−10.4227	−3.4476	893 (0.0262)	(0.1601, 0.2273)	0.0908
True value	1.1365	0.1036	−0.0894	1.3159	−2.6000	5.4000	—	—	—
FH-15(-2)	1.1387	0.0984	−0.0736	1.3238	−2.2175	1.7098	47 (0.0408)	(0.7131, 0.9156)	3.7101
FH-9(-1)	1.1402	0.1055	−0.0805	1.3160	−3.6578	3.5156	29 (0.1212)	(1.0153, 0.9129)	2.1610
FH-9(-2)	1.1361	0.1038	−0.0897	1.3160	−2.3833	5.6308	157 (0.0427)	(0.3856, 0.4829)	0.3166
FH-9(-3)	1.1363	0.1038	−0.0895	1.3165	−2.4408	5.4805	476 (0.0206)	(0.1902, 0.3197)	0.1784
FH-9(-4)	1.1365	0.1037	−0.0894	1.3159	−2.4575	5.5336	983 (0.0180)	(0.1616, 0.2378)	0.1954
FH-9(-5)	1.1363	0.1037	−0.0894	1.3160	−2.4582	5.5270	1293 (0.0300)	(0.1432, 0.2119)	0.1903

Detectors	Estimated affine warping parameters						Correct match number and *MFAR*	*ATE*	*WMEE*
	a	b	c	d	t_x	t_y			
True value	1.2079	0.0777	−0.0718	1.3077	−5.3000	1.5000	—	—	—
FH-15(-2)	1.2033	0.0744	−0.0753	1.3054	−4.2454	2.4055	52 (0.0545)	(0.7247, 0.7616)	1.3900
FH-9(-1)	1.1959	0.0695	−0.0650	1.3079	−1.8954	0.0939	24 (0.0769)	(0.9570, 1.1486)	3.6836
FH-9(-2)	1.2075	0.0778	−0.0735	1.3066	−5.0414	2.0074	172 (0.0227)	(0.3858, 0.4182)	0.5695
FH-9(-3)	1.2076	0.0766	−0.0719	1.3077	−5.0751	1.6740	514 (0.0172)	(0.2207, 0.3552)	0.2844
FH-9(-4)	1.2078	0.0777	−0.0718	1.3077	−5.1181	1.6590	1052 (0.0177)	(0.1506, 0.2289)	0.2416
FH-9(-5)	1.2077	0.0778	−0.0719	1.3078	−5.1297	1.6397	1451 (0.0176)	(0.1343, 0.1998)	0.2203

Table 2.
Evaluation of SURF Fast-Hessian detectors on four controlled SAR image pairs.

4. Appropriate retrieval algorithm for SAR image registration

The next procedure after feature extraction is to retrieve the warp function from the attained correspondences. Due to the influences of spatial/temporal decorrelation, system noise, and environmental interference, or the non-robustness in the depiction and matching of features, there are always mismatches in the constructed correspondences. It is difficult to get *a priori* information to remove them beforehand. To accurately retrieve parameters from these error-prone correspondences, some robust outlier-insensitive algorithms are necessary.

Furthermore, unlike the pinhole imaging of optical camera, SAR acquires the imagery using a slant-range geometry which cannot be modeled as a central projection [47]. As a result, the warp model between SAR images is dependent on the system parameter, imaging geometry, and target relief, and we cannot adopt a global homography or essential matrix to model the geometrical warping then. Nevertheless, when the system parameter and imaging geometry are fixed and the area-of-interest has gentle topography, we can conventionally approximate the warp function as a low-order polynomial [48]. This indicates our strategy in the retrieval of registration parameters, to focus on the global registration instead of local discontentment.

4.1 Evaluation of RANSAC for SAR image registration

RANSAC [30] has been widely used in feature-based SAR image registrations for parameter retrieval [15, 16, 26, 27]. Unlike LS which uses all the available data to estimate parameters, RANSAC conducts the estimation using a few-to-many strategy or a local-to-global strategy. A MSS is randomly sampled from the constructed correspondences to achieve an estimation of the warp function firstly. The cardinality of MSS, i.e., the smallest sufficiency to determine the warp parameters, is just related to the degree of freedom (DoF) of the warp function. For example, the cardinality will be 3 for affine transformation of 6 DOFs. The entire dataset are then checked for those correspondences consistent with the retrieved warping to construct a larger CS. These two steps are repeated until the largest CS is finally achieved for parameter estimation. This local-to-global strategy is tenable only if any MSS of inliers can generate the "true value" of warp parameters [31]. But it is often hard to keep this in real registration due to the unavoidable noise and local distortion, i.e., a different estimation of parameters will be achieved from a different MSS configuration of inliers. This uncertainty is even more severe in SAR image registration because SAR warping varies from pixel to pixel and the low-order polynomial approximation only accounts for global registration instead of local contentment. The local-to-global strategy may then magnify the local distortion, aggravate the estimation uncertainty, and damnify the global registration accuracy although a largest CS is identified. To demonstrate this, we devise an experiment to coregister a spaceborne InSAR image pair as shown in **Figure 4(a)** and **(b)**. The two images are acquired by RadarSat-2 on May 4 and 28, 2008, respectively. The scene is within South Phoenix, AZ, USA with some buildings and vegetable lands. We first use *FH*-9(-1) to construct SURF feature correspondences, and then adopt RANSAC to retrieve the affine warp parameters. To evaluate the estimation certainty, we execute RANSAC 100 times and based on the obtained parameters of each execution, we coregister the complex image pair to calculate the three-look coherent CC and spectral SNR. CC measures the consistency, while spectral SNR, the ratio between the maximum entry and the sum of other entries in the spectrum, reflects the clarity of the interferogram fringe [49]. **Figure 5** displays the affine parameters a, b, c, d, t_x, and t_y as well as CC and SNR obtained in each execution. **Table 3** further displays the mean and standard deviation of the parameters, CC, and SNR. RANSAC cannot obtain a stable registration because the retrieval parameters vary with executions, even for executions with the same cardinality of CS achieved. **Figure 6** shows the retrieval parameters, CC and SNR for 48 executions with the same cardinality. We can still find the estimation uncertainty. This reveals that the attained inliers which compose the final CS are actually different although the same cardinality. Otherwise, the parameters would be the same for each execution because they are retrieved by just LS fitting the inliers.

The uncertainty of RANSAC in SAR image registration just comes from its retrieval strategy and loss function. To achieve a stable registration for SAR images, a feasible improvement is to estimate the parameters with more correspondences to reflect the true support than just a MSS, and to apply an appropriate loss function. This leads us another direction to the robust parameter regression.

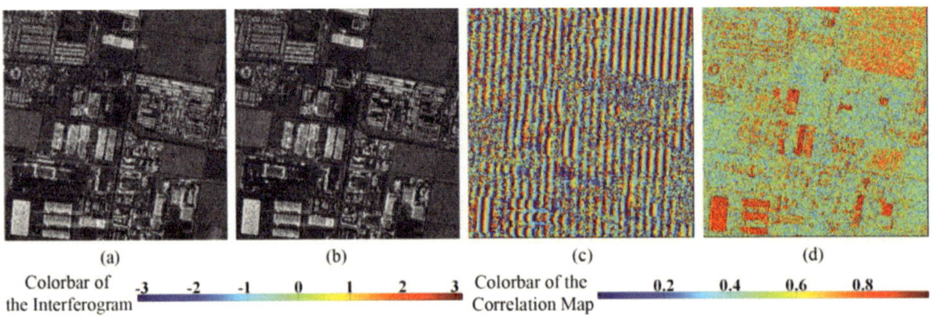

Figure 4.
Registration of InSAR image pair from RadarSat-2. (a) Master image, (b) slave image, and the final (c) interferogram and (d) correlation map based on the registration parameters estimated by EF-LTS.

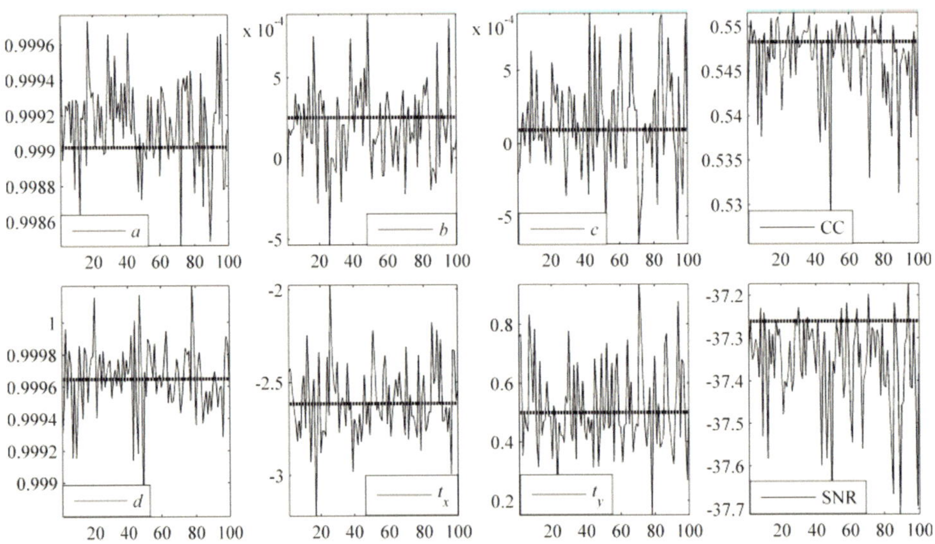

Figure 5.
Registration parameters, cross-correlation (CC), and spectral SNR obtained by 100 executions of (thin line) RANSAC and (thick line) EF-LTS on the image pair of RadarSat-2.

Algorithm	a		b		c		CC	
	Mean	Std	Mean	Std	Mean	Std	Mean	Std
RANSAC	0.9992	2.5228×10^{-4}	1.8796×10^{-4}	2.4879×10^{-4}	1.3770×10^{-4}	3.3408×10^{-4}	0.5462	0.0046
EF-LTS	0.9990	0.0000	2.5485×10^{-4}	0.0000	9.2440×10^{-5}	0.0000	0.5483	0.0000
Algorithm	d		t_x		t_y		SNR (dB)	
	Mean	Std	Mean	Std	Mean	Std	Mean	Std
RANSAC	0.9996	2.2264×10^{-4}	−2.6068	0.1955	0.5133	0.1509	−37.36	0.1126
EF-LTS	0.9996	0.0000	−2.6151	0.0000	0.5008	0.0000	−37.26	0.0000

Table 3.
Mean and standard deviation of the registration parameters, cross-correlation (CC), and spectral SNR obtained by RANSAC and EF-LTS on RadarSat-2 InSAR images.

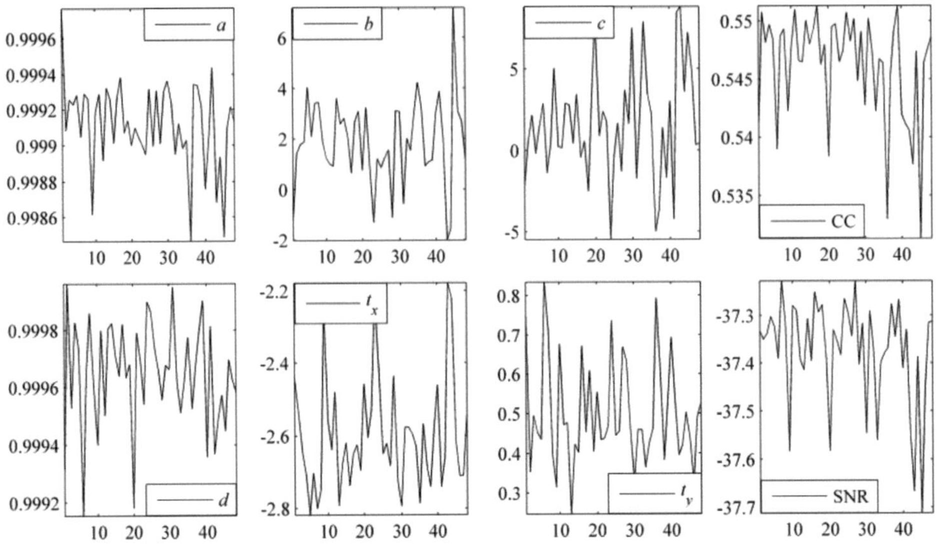

Figure 6.
Registration parameters, cross-correlation (CC), and spectral SNR obtained by 48 executions of RANSAC on RadarSat-2 InSAR images with the same CS cardinality.

4.2 Fast-LTS

The widely used LS is now being criticized more and more for lack of robustness. To tackle with this, some robust regression approaches were developed, like LMedS [32] and the least trimmed squares (LTS) [50]. LMedS implements the regression by minimizing the median of residual squares. This makes LMedS so robust that it can still obtain a reasonable estimation even if 50% of the dataset are outliers. So the breakdown point of LMedS is as high as 50%. LTS is a modification of LS with the same breakpoint as LMedS. It also fits the linear model:

$$y_i = \mathbf{X}_i^{\mathrm{T}}\boldsymbol{\theta} + e_i, \quad i = 1, ..., n \tag{19}$$

where $\mathbf{X}_i = [x_{i1}, x_{i2}, ..., x_{ip}]^{\mathrm{T}}$ denotes the explanatory variable, y_i denotes the response variable, $\boldsymbol{\theta} = [\theta_1, \theta_2, ..., \theta_p]^{\mathrm{T}}$ indicates the unknown parameter to be retrieved, e_i is the error term, n is the sample size, and p is the dimension of $\mathbf{X}_i$. The loss function of LTS is:

$$Q{:=}\text{Minimize} \sum_{i=1}^{h} (\mathbf{r}^2)_i \text{ with } \mathbf{r} = [r_1, r_2, \cdots, r_n]^{\mathrm{T}} \text{ and } r_i = y_i - \mathbf{X}_i^{\mathrm{T}}\boldsymbol{\theta} \tag{20}$$

where $(\mathbf{r}^2)_i$ denotes the ith element of the ordered squared residuals $(\mathbf{r}^2)_1 \le \cdots \le (\mathbf{r}^2)_i \le \cdots \le (\mathbf{r}^2)_n$, and h is termed as the trimming constant. LTS conducts regression by LS fitting the h-subset to minimize the squared residuals. Compared with LMedS, the statistical efficiency of LTS is much better and the loss function is much smoother [33]. Nevertheless, the deficiency of LTS is the large computation when processing the big data. To accelerate it, Rousseeuw and Van Driessen [33] developed a Fast-LTS, which can efficiently deal with a sample size as large as tens of thousands or even larger. The core of Fast-LTS is a concentration step (C-step), which is designed to achieve a better estimation from an old h-subset $\mathbf{H}_{\mathbf{old}}$ [33]:

Algorithm 1: C-step

Step 11. Compute regression parameters $\boldsymbol{\theta}_{\mathbf{old}}$ by LS fitting $\mathbf{H}_{\mathbf{old}}$.

Step 12. Calculate residuals $\mathbf{r}_{\mathbf{old}}$ based on $\boldsymbol{\theta}_{\mathbf{old}}$. Ascendingly sort squared residuals $\mathbf{r}_{\mathbf{old}}{}^2$ for a permutation $\boldsymbol{\pi}$ of the set such that $(\mathbf{r}_{\mathbf{old}}{}^2)_{\pi(1)} \leq \cdots \leq (\mathbf{r}_{\mathbf{old}}{}^2)_{\pi(h)} \leq \cdots \leq (\mathbf{r}_{\mathbf{old}}{}^2)_{\pi(n)}$.

Step 13. Construct a new h-subset $\mathbf{H}_{\mathbf{new}} = \{\boldsymbol{\pi}(1), \boldsymbol{\pi}(2), \ldots, \boldsymbol{\pi}(h)\}$ and obtain the new parameters $\boldsymbol{\theta}_{\mathbf{new}}$ by LS fitting $\mathbf{H}_{\mathbf{new}}$.

It has been proved that Q of parameters $\boldsymbol{\theta}_{\mathbf{new}}$ is always no larger than that of parameters $\boldsymbol{\theta}_{\mathbf{old}}$ [33]. Therefore, an improved estimation of parameters can be achieved after an execution of C-step, and a converged Q will be obtained after only a few C-steps. Thus Fast-LTS conducts estimation as follows [33]:

Algorithm 2: Fast-LTS

Step 21. Randomly generate a p-subset as parameter set $\boldsymbol{\theta}_{\mathbf{0}}$. Calculate n residuals $\mathbf{r}_{\mathbf{0}}$ based on $\boldsymbol{\theta}_{\mathbf{0}}$ to achieve an initial h-subset $\mathbf{H}_{\mathbf{0}} = \{\boldsymbol{\pi}(1), \boldsymbol{\pi}(2), \ldots, \boldsymbol{\pi}(h)\}$ such that $(\mathbf{r}_{\mathbf{0}}{}^2)_{\pi(1)} \leq \cdots \leq (\mathbf{r}_{\mathbf{0}}{}^2)_{\pi(h)} \leq \cdots \leq (\mathbf{r}_{\mathbf{0}}{}^2)_{\pi(n)}$. Update $\mathbf{H}_{\mathbf{0}}$ by carrying out two C-steps on $\mathbf{H}_{\mathbf{0}}$. Repeat above procedures 500 times.

Step 22. Implement C-steps on the 10 $\mathbf{H}_{\mathbf{0}}$ with the lowest 10 Q until convergence. Then the solution that creates the lowest Q is identified as the final estimation $\boldsymbol{\theta}$.

The trimming constant h is set between $[(n + p + 1)/2)$ ($[x)$ denotes the smallest integer larger than x) and n. The breakdown value of Fast-LTS is $(n - h + 1)/n$. A nested extension approach should be adopted to enable an efficient estimation when n is larger [33].

4.3 EF-LTS for SAR image registration

Fast-LTS is appropriate for 1D linear regression formulated in (19). However, for SAR image registration, what we need to do is to fit a 2D polynomial regression

$$\begin{cases} x_{si} = \sum\limits_{j=0}^{N} \sum\limits_{k=0}^{N-j} a_{jk} x_{mi}^{j} y_{mi}^{k} + \varsigma_i \\ y_{si} = \sum\limits_{j=0}^{N} \sum\limits_{k=0}^{N-j} b_{jk} x_{mi}^{j} y_{mi}^{k} + \xi_i \end{cases}, \quad i = 1, \ldots, n \tag{21}$$

where n is the number of constructed correspondences, N is the order of polynomial, a and b are polynomial coefficients, (x_{si}, y_{si}) and (x_{mi}, y_{mi}) are the ith feature correspondence extracted from the slave and master images, and ζ_i and ξ_i denote the normally distributed error terms with zero mean. Actually, (21) denote a 2D linear regression problem:

$$\begin{cases} x_{si} = \mathbf{X}_i^{\mathrm{T}} \boldsymbol{\theta} + \varsigma_i \\ y_{si} = \mathbf{X}_i^{\mathrm{T}} \boldsymbol{\psi} + \xi_i \end{cases} \text{with} \begin{cases} \boldsymbol{\theta} = [\theta_1, \theta_2, \cdots, \theta_p]^{\mathrm{T}} = [a_{00}, a_{01}, \cdots, a_{N0}]^{\mathrm{T}} \\ \boldsymbol{\psi} = [\psi_1, \psi_2, \cdots, \psi_p]^{\mathrm{T}} = [b_{00}, b_{01}, \cdots, b_{N0}]^{\mathrm{T}} \\ \mathbf{X}_i = [X_{i1}, X_{i2}, \cdots, X_{ip}]^{\mathrm{T}} = [1, y_{mi}, \cdots, y_{mi}^{N}, x_{mi}, \cdots, x_{mi}^{N}]^{\mathrm{T}} \end{cases}, \quad i = 1, \ldots, n \tag{22}$$

where $\boldsymbol{\theta}$ and $\boldsymbol{\psi}$ are the unknown parameters to be estimated, and $p = (N + 1)(N + 2)/2$ denotes the number of unknowns. Then, the warp function estimation for SAR image registration can be transformed into the following optimization problems:

$$\begin{cases} Q_x\text{:=Minimize} \sum_{i=1}^{h} (\mathbf{r}_\mathbf{x}^2)_i \\ Q_y\text{:=Minimize} \sum_{i=1}^{h} \left(\mathbf{r}_\mathbf{y}^2\right)_i \end{cases} \text{with} \begin{cases} \mathbf{r}_\mathbf{x} = [r_{x1}, r_{x2}, \cdots, r_{xn}]^\mathrm{T} \\ \mathbf{r}_\mathbf{y} = [r_{y1}, r_{y2}, \cdots, r_{yn}]^\mathrm{T} \end{cases} \text{and} \begin{cases} r_{xi} = x_{si} - \mathbf{X}_i^\mathrm{T}\boldsymbol{\theta} \\ r_{yi} = y_{si} - \mathbf{X}_i^\mathrm{T}\boldsymbol{\psi} \end{cases}, \quad i = 1, \ldots, n \tag{23}$$

where $(\mathbf{r_x}^2)_i$ represents the ith element of the ordered squared residuals $(\mathbf{r_x}^2)_1 \le \cdots \le (\mathbf{r_x}^2)_i \le \cdots \le (\mathbf{r_x}^2)_n$, and the meaning of $(\mathbf{r_y}^2)_i$ can be likewise inferred. Each of the two optimizations in (23) is of the standard form (20). A direct solution to (23) may be thus achieved by decomposing 2D regression as two independent 1D regressions and using Fast-LTS to conduct estimation, respectively. This idea is feasible, but it may result in unnecessary computations because the feature positions in two image directions are in fact tied to each other, i.e., for the ith feature (x_i, y_i), the selection of x_i will naturally mean the selection of y_i. We can thus combine the two 1D regressions into a real 2D regression effectively, i.e., the extended Fast-LTS (EF-LTS):

Algorithm 3: EF-LTS

Step 31. Randomly draw p feature matches and LS fit them to estimate the initial parameters $\boldsymbol{\theta}_\mathbf{0}$ and $\boldsymbol{\psi}_\mathbf{0}$, and calculate the initial residuals $\mathbf{r_{0x}}$ and $\mathbf{r_{0y}}$ by

$$\begin{cases} \mathbf{r_{0x}} = [r_{0x1}, r_{0x2}, \cdots, r_{0xn}]^\mathrm{T} \\ \mathbf{r_{0y}} = [r_{0y1}, r_{0y2}, \cdots, r_{0yn}]^\mathrm{T} \end{cases} \text{and} \begin{cases} r_{0xi} = x_{si} - \mathbf{X}_i^\mathrm{T}\boldsymbol{\theta}_\mathbf{0} \\ r_{0yi} = y_{si} - \mathbf{X}_i^\mathrm{T}\boldsymbol{\psi}_\mathbf{0} \end{cases}, \quad i = 1, \ldots, n. \tag{24}$$

Then construct the initial h-subsets $\mathbf{Hx_0}$ and $\mathbf{Hy_0}$ by:

$$\begin{cases} \mathbf{Hx_0} = \{\boldsymbol{\pi}\mathbf{x}(1), \boldsymbol{\pi}\mathbf{x}(2), \cdots, \boldsymbol{\pi}\mathbf{x}(h)\} \subset \{1, 2, \cdots, n\} \\ \mathbf{Hy_0} = \{\boldsymbol{\pi}\mathbf{y}(1), \boldsymbol{\pi}\mathbf{y}(2), \cdots, \boldsymbol{\pi}\mathbf{y}(h)\} \subset \{1, 2, \cdots, n\} \end{cases} s.t. \begin{cases} \left(\mathbf{r_{0x}^2}\right)_{\boldsymbol{\pi}\mathbf{x}(1)} \le \cdots \le \left(\mathbf{r_{0x}^2}\right)_{\boldsymbol{\pi}\mathbf{x}(h)} \le \cdots \le \left(\mathbf{r_{0x}^2}\right)_{\boldsymbol{\pi}\mathbf{x}(n)} \\ \left(\mathbf{r_{0y}^2}\right)_{\boldsymbol{\pi}\mathbf{y}(1)} \le \cdots \le \left(\mathbf{r_{0y}^2}\right)_{\boldsymbol{\pi}\mathbf{y}(h)} \le \cdots \le \left(\mathbf{r_{0y}^2}\right)_{\boldsymbol{\pi}\mathbf{y}(n)} \end{cases}. \tag{25}$$

Carry out two C-steps on $\mathbf{Hx_0}$ and $\mathbf{Hy_0}$ to obtain the h-subsets $\mathbf{Hx_2}$ and $\mathbf{Hy_2}$ with smaller Q_x and Q_y, respectively. Iteratively repeat above procedures T times to obtain a set of h-subsets $\mathbf{Hx_2}$ and $\mathbf{Hy_2}$.

Step 32. Select 10 $\mathbf{Hx_2}$ with the smallest 10 Q_x and 10 $\mathbf{Hy_2}$ with the smallest 10 Q_y if T is larger than 10; otherwise, select all $\mathbf{Hx_2}$ and $\mathbf{Hy_2}$. Carry out C-steps on these h-subsets until convergence. The solutions corresponding to the smallest Q_x and Q_y are selected as the raw estimations $\boldsymbol{\theta}_\mathbf{r}$ and $\boldsymbol{\psi}_\mathbf{r}$, respectively.

Step 33. Calculate residuals $\mathbf{r_{rx}}$ and $\mathbf{r_{ry}}$ based on $\boldsymbol{\theta}_\mathbf{r}$ and $\boldsymbol{\psi}_\mathbf{r}$,

$$\begin{cases} \mathbf{r_{rx}} = [r_{rx1}, r_{rx2}, \cdots, r_{rxn}]^\mathrm{T} \\ \mathbf{r_{ry}} = [r_{ry1}, r_{ry2}, \cdots, r_{ryn}]^\mathrm{T} \end{cases} \text{with} \begin{cases} r_{rxi} = x_{si} - \mathbf{X}_i^\mathrm{T}\boldsymbol{\theta}_\mathbf{r} \\ r_{ryi} = y_{si} - \mathbf{X}_i^\mathrm{T}\boldsymbol{\psi}_\mathbf{r} \end{cases}, \quad i = 1, \ldots, n \tag{26}$$

and estimate the error scales σ_x and σ_y by

$$\begin{cases} \sigma_x = C_1\sqrt{\dfrac{1}{h}\sum_{i=1}^{h}\left(\mathbf{r_{rx}^2}\right)_i} \\ \sigma_y = C_2\sqrt{\dfrac{1}{h}\sum_{i=1}^{h}\left(\mathbf{r_{ry}^2}\right)_i} \end{cases} \tag{27}$$

where C_1 and C_2 are correction factors to achieve consistency at Gaussian error distributions [50]. Based on (27), we further calculate two weights by:

$$w_{xi} = \begin{cases} 1 & \text{if } |r_{xi}/\sigma_x| \le 2.5 \\ 0 & \text{if } |r_{xi}/\sigma_x| > 2.5 \end{cases} \quad w_{yi} = \begin{cases} 1 & \text{if } |r_{yi}/\sigma_y| \le 2.5 \\ 0 & \text{if } |r_{yi}/\sigma_y| > 2.5 \end{cases}, \quad i = 1, \ldots, n. \tag{28}$$

The credible correspondence in both directions of x and y is chosen:

$$w_i = w_{xi} \,\&\, w_{yi}, \quad i = 1, \ldots, n \tag{29}$$

where "&" denotes the logical AND operator. The final estimations $\boldsymbol{\theta}_\mathbf{f}$ and $\boldsymbol{\psi}_\mathbf{f}$ are attained by LS solving the following optimizations:

$$\begin{cases} \boldsymbol{\theta}_\mathbf{f} = \operatorname{argmin} \sum_{i=1}^{n} w_i r_{xi}^2 \\ \boldsymbol{\psi}_\mathbf{f} = \operatorname{argmin} \sum_{i=1}^{n} w_i r_{yi}^2 \end{cases} \tag{30}$$

which in fact indicates the weighted LS.

Step 33 makes EF-LTS obtain more accurate and stable estimation than the original LTS. The logical AND in (29) shows that only the feature correspondence which is correctly matched in both x- and y-direction is considered as an inlier. This is necessary for accurate estimation because mismatching in one direction may also affect the matching in another. The bound in (28) is set as 2.5 for there are very few residuals larger than 2.5σ in a Gaussian situation [50].

In Fast-LTS, the random sampling number T is a constant 500. This is inappropriate because accurate estimation only requires one p-match to being "clean." Let q denote the percentage of inliers in data, then the probability ε of having at least one "clean" p-match among all the T random p-matches can be expressed as

$$\varepsilon = 1 - (1 - q^p)^T. \tag{31}$$

Since the trimming constant h is chosen beforehand according to the percentage of inliers, a good estimation of q can be obtained by

$$\hat{q} = \frac{h}{n}. \tag{32}$$

Therefore, if a required false alarm rate ε for the estimation is given, the sampling number T can be then calculated by combining (31) and (32):

$$T = \left\lceil \frac{\log(1 - \varepsilon)}{\log\left(1 - \left(\frac{h}{n}\right)^p\right)} \right). \tag{33}$$

Thus, iteration in EF-LTS is controlled by the inlier percentage rather than the inlier number. **Table 4** shows the sampling number T under given N and q when $\varepsilon = 0.99$. It can be seen that even the worst sampling number 293 is much smaller than 500 for $N = 2$. Thus, the constant 500 sampling will be redundant for the second-order polynomial, but will be insufficient for the third-order polynomial with smaller q, as listed in **Table 4**.

The inlier percentage q is in fact related to $MFAR$ by:

$$q = 1 - MFAR. \tag{34}$$

Thus, besides introducing more iterations and computation load, higher $MFAR$ will also lead to a smaller h-subset, which indicates more localization and less

N	*q*						
	0.5	0.6	0.7	0.75	0.8	0.9	0.95
0	7	6	4	4	3	2	2
1	35	19	11	9	7	4	3
2	293	97	37	24	16	7	4
3	4714	760	161	80	41	11	6

Table 4.
Sampling number T under different inlier percentage q and polynomial order N when $\varepsilon = 0.99$.

accuracy in estimation and worse consistency between the extracted features and retrieval parameters. This is why *FH*-15(-2) can achieve better *ATE* than *FH*-9(-1), as displayed in **Table 2**. As presented in Section 3, on *MFAR* and many other criteria, SURF is identified to be the best for general SAR image registration. SURF may thus also improve the efficiency and accuracy of parameter retrieval besides the good performance on feature extraction and matching.

When the correspondence number n is large, a similar nested extension can be also taken for EF-LTS by randomly partitioning the correspondences into M subsets with equal cardinality, and the trimming constant h_s and sampling number T_s of each subset should be also reduced by M times relative to h and T. On each subset, we first implement Step 31 for T_s h_s-subsets of $\mathbf{Hx_2}$ and $\mathbf{Hy_2}$. Based on which we then implement Step 32 and Step 33 on all the constructed correspondences with original h and T. In this way, an efficient retrieval can be still achieved.

To evaluate EF-LTS for SAR image registration, we also use it to the InSAR image pair given in **Figure 4(a)** and **(b)**. Similarly, the feature correspondences are first constructed by SURF with *HF*-9(-1), then we run EF-LTS 100 times to retrieve the affine parameters and calculate CC and SNR. The obtained parameters, CC, and SNR of each execution are shown in **Figure 5**, while the mean and standard deviation of the parameters, CC, and SNR are listed in **Table 3**. It is revealed that EF-LTS behaves very stable and the estimated parameters, CC, and SNR are invariant for each execution. It can reach an averagely better CC and SNR than RANSAC and is more appropriate for InSAR image registration. **Figure 4(c)** and **(d)** further illustrates the interferogram and correlation map of the coregistered InSAR pair with wrap parameters estimated by EF-LTS. Interferogram is the argument or phase of the dot production between the complex master image and the complex conjugation of the registered slave image, while correlation map measures CC of the 3×3 patches around each corresponding pixel position between the images. The interferogram fringe is clear and the correlation is strong in stable area such as the brighter buildings in **Figure 4(a)** and **(b)** and the upper-right bare land. But in the upper-left residential area, the interferogram becomes less clear and the correlation is relatively small probably because the scattering is very sensitive to incidence changes. While in other area (mainly vegetable lands and parking lot), the interferogram is almost lost and the coherence is very low due to the temporal and/or volume decorrelation. All these match with the ground truth very well.

5. Experiment and analysis

Based on the finding in Sections 2–4, we propose to conduct high accurate SAR image registration by using EF-LTS to fit the SURF correspondences. The scheme works as follows:

Algorithm 4: Accurate SAR image registration based on SURF features and EF-LTS

Step 41. Use *FH*-9(-*Fs*) to extract SURF keypoints from master and slave images, respectively.

Step 42. Construct initial feature correspondences by simply matching SURF descriptors.

Step 43. Robustly processing the correspondences with EF-LTS to retrieve the warp function.

Step 44. Transform and interpolate the slave image to geometrically align it to master image.

Actually, this scheme has been put into practice in the above experiments. In this section, we further devise an experiment to check it on MiniSAR pair. The images we use are two high-resolution SAR images of the entrance gate of the Sandia Research Park acquired by the Ku-Band MiniSAR system developed by the Sandia Laboratory [51]. The images are taken from different tracks with different incidences and squints, as listed in **Table 5**, while the platform altitude is just beyond 1 km. All these reveal the nontrivial target relief-induced geometrical warping between images, which, however, cannot be compensated beforehand for lack of ground truth such as DEM and target height. Besides this, the images also experience a very large intensity variation. To enhance the texture, we use the logarithmic intensity of original complex images, as shown in **Figure 7(a)** and **(b)**. To achieve a more precise approximation to the real warping, we divide the image pair into four 500×500 patch pairs. The geometrical warping on each patch pair is approximated as an affine transformation (the higher order polynomial has also been used to model the warp function, but unsatisfactory registration result is attained). We adopt *HF*-9(-4) SURF detector to extract feature correspondences from each patch pair, and EF-LTS is then used to obtain the affine parameters, based on which the slave image is finally aligned to the master image. To illustrate the registration accuracy, we fuse and overlap the coregistered images together. The RGB fusion in **Figure 7(c)** is obtained by treating the master image and the coregistered slave image as red and green, respectively, while zeroing the blue component. The well-distributed yellow then immediately illustrates the accurate registration of the images. The overlapping in **Figure 7(d)** is obtained by simply averaging the two coregistered images. It contains the whole information of the two images but has fewer speckles.

To further evaluate the registration performance of the scheme, in the following we focus on the two pole-like target areas 1 and 2 in **Figure 7(d)** with their corresponding Google optical images shown in **Figure 8(g)** and **(h)**, respectively. **Figure 8(i)** portrays the details of Pole 2 in the Street View of Google Maps. The

Parameters	Master image	Slave image
Azimuth resolution	0.1016 m	0.1016 m
Range resolution	0.1016 m	0.1016 m
Grazing angle	27.0107°	26.1892°
Global track angle	158.3687°	153.0825°
Central frequency	16.8 GHz	16.8 GHz
Platform altitude	1.6715 km	1.6715 km
Squint	−89.9935°	−89.9924°

Table 5.
Imaging parameters of the two MiniSAR images.

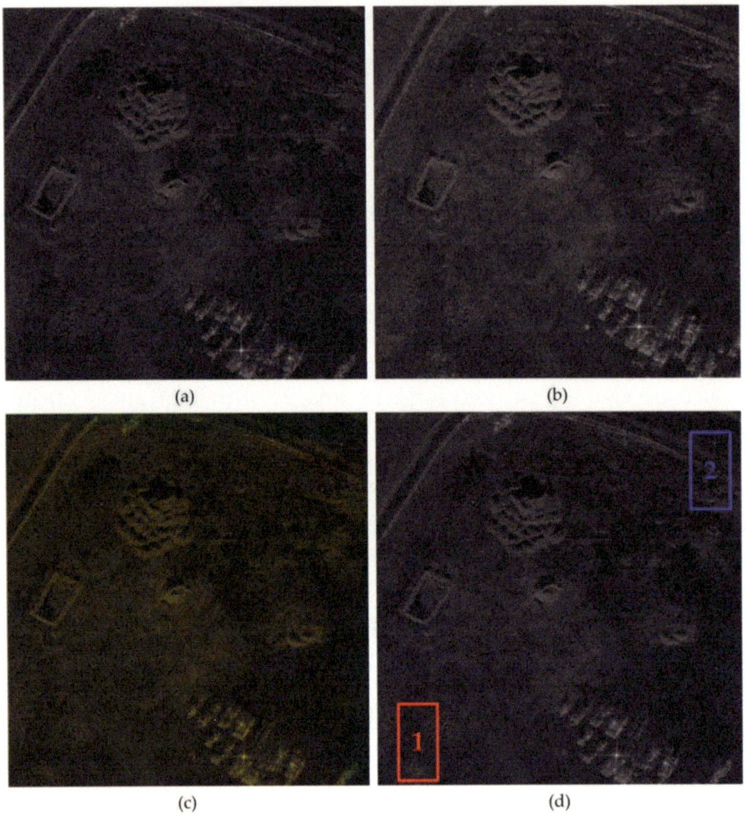

Figure 7.
Registration of the MiniSAR image pair. (a) Master image, (b) slave image, and (c) pseudocolor fusion as well as (d) the overlapping of them after registration with EF-LTS to fit the SURF features. "1" and "2" in (d) indicate two pole-like targets which are further detailed in ***Figure 8****.*

target is shown to be the power transmission pole. **Figure 8(a)–(c)** exhibits the SAR imagery of Pole 1 in the master image, coregistered slave image, and overlapped image, respectively. The corresponding SAR imageries of Pole 2 are displayed in **Figure 8(d)–(f)**, respectively. It is known that the darker pole-like feature in each SAR image is not the real pole scattering, but its shadow under the irradiation of radar. The actual scattering center of the pole is overlapped with its ground position because of the dominant dihedral backscattering between the pole and ground. From **Figure 8(c)** and **(f)**, we can find that the shadows of the two poles are still separated after registration due to the volume-induced warping. According to our estimate, the separations are about 6.5 and 5°, respectively, which approach to the actual track angle 5.2862°. Except for these shadows, the poles and other area are accurately overlapped. Nice registration is still achieved despite the large local distortion and decorrelation. Moreover, the experiment also validates the strategy for general feature-based SAR image registration, i.e., to focus on the global registration and to neglect the local discontentment. The accurate registration of each pixel is impossible and unnecessary. It should be noted that the conventional SAR image registrations including the feature-based approaches focused in current chapter are mainly appropriate for images with approximated low-order polynomial geometrical warping. For SAR images taken from area of rough topography with long baseline, we need some more complex approaches with the *a priori* ground truth information being included, such as the DEM-assisted registration [48]. Although the SAR and InSAR image pairs used in the experiment are all

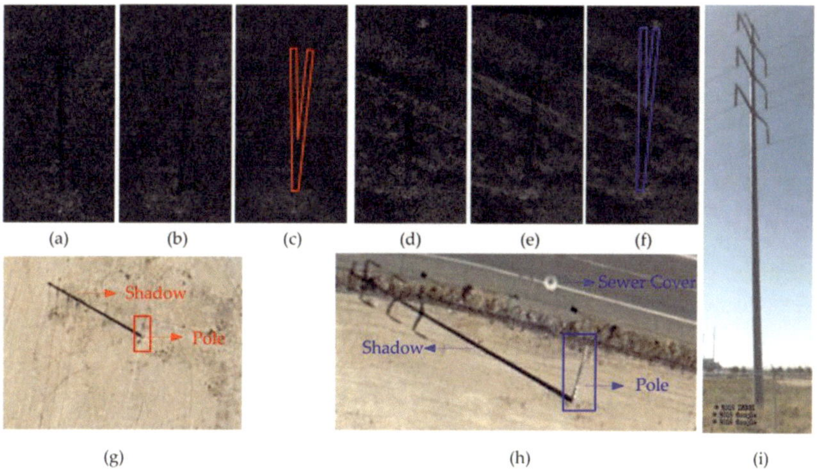

Figure 8.
Registration of the two pole-like targets. SAR imagery of pole "1" in (a) master image, (b) coregistered slave image, and (c) overlapped image, as well as (g) the corresponding Google optical image. SAR imagery of pole "2" in (d) master image, (e) coregistered slave image, and (f) overlapped image, as well as (h) its Google optical image and (i) detailed portrayal in Google Maps.

monopolarized, the developed scheme is also appropriate to the registration of fully polarimetric SAR (PolSAR) images. Different from monopolarized SAR, each cell in PolSAR image is a scattering matrix **S** with four entries S_{HH}, S_{HV}, S_{VH}, and S_{VV} [52]:

$$\mathbf{S} = \begin{bmatrix} S_{HH} & S_{HV} \\ S_{VH} & S_{VV} \end{bmatrix}. \tag{35}$$

Nevertheless, by taking the squared Frobenius norm of matrix **S** [53]:

$$SPAN = \|\mathbf{S}\|_F^2 = |S_{HH}|^2 + |S_{HV}|^2 + |S_{VH}|^2 + |S_{VV}|^2 \tag{36}$$

we can then obtain the total power (also known as *SPAN*) of target. An accurate registration of PolSAR images can be eventually achieved by simply using the developed scheme to the corresponding *SPAN* image pair.

6. Conclusion

SAR coherent imaging unavoidably brings about geometrical distortion and speckle into the acquired images and makes the registration of SAR images much more complicated. In this chapter, we focus on two important procedures in general feature-based SAR registration, i.e., the feature extraction and the parameter retrieval by identifying the appropriate feature and the appropriate estimation algorithm. As for the former, we conduct a detailed evaluation on the commonly used features such as tie points, Harris corner, SIFT, and SURF. We find that SURF outperforms others in terms of the geometrical invariance of feature, extraction speed, accuracy of localization, geometrical invariance of descriptor, matching speed, robustness to decorrelation, and flexibility to image speckling. Among these criteria, feature's flexibility to speckle is particularly focused because speckle impacts the feature extraction and matching, while speckle filtering may change the feature position and impact the subpixel localization. The Fast-Hessian detector of

SURF has a potential relation with the refined Lee speckle filter. SSP in SURF just indicates that we use a series of box filters of different size to filter speckles and extract features of different scales. Thus, SURF is very flexible to deal with SAR speckle. In view of the application with strict requirement for registration accuracy, we suggest using the SURF detector of *HF*-9(-1) to the *Fs* times interpolated images with unchanged sampling step to extract feature. The new detector *HF*-9(-*Fs*) can significantly improve the registration accuracy to subpixel (<1 pixel) and is especially fit for high accurate SAR image registration.

Parameter retrieval in SAR registration is difficult because spatial or temporal decorrelation will always introduce mismatches into the obtained feature correspondences. The estimator should be robust to outliers. We find that the commonly used RANSAC may trap into local occlusion and result in uncertain parameter retrieval. This uncertainty is more severe in SAR image registration because SAR geometrical warping varies from pixel to pixel, but the low-order polynomial approximation can only account for global registration instead of the local contentment. The local-to-global strategy in RANSAC may thus magnify the local distortion, aggravate the estimation uncertainty, and damnify the global registration accuracy although a largest CS is obtained. To achieve a stable registration for SAR images, we should estimate the parameters with more correspondences to reflect the true support than just a MSS, and apply an appropriate loss function. This leads us to EF-LTS, which improves Fast-LTS from 1D regression to 2D regression, and provides us an adaptive determination of the number of random sampling instead of setting it as a constant 500. EF-LTS conducts registration by LS fitting at least half of the correspondences to minimize the squared residual. It behaves very stable and is averagely better than RANSAC. Hence, we recommend conducting SAR image registration by fitting SURF features with EF-LTS. Experiments on both InSAR and MiniSAR image pairs validate the nice performance of this registration scheme.

Acknowledgements

This work is supported by China Manned Space Program along with the Youth Innovation Promotion Association, Chinese Academy of Sciences under Grant No. 2014131. The authors thank the International Society for Optics and Photonics (SPIE) for the permission to reuse materials that have appeared in Proceedings of SPIE (Li D, Zhang Y. On the appropriate feature for general SAR image registration; The appropriate parameter retrieval algorithm for feature-based SAR image registration. SAR Image Analysis Modeling and Techniques XII. Vol. 8536, 2012.)

Author details

Dong Li[1], Yunhua Zhang[1,2]* and Xiaojin Shi[1]

1 Key Laboratory of Microwave Remote Sensing, National Space Science Center, Chinese Academy of Sciences, Beijing, China

2 University of Chinese Academy of Sciences, Beijing, China

*Address all correspondence to: zhangyunhua@mirslab.cn

References

[1] Homer J, Longstaff ID. Minimising the tie patch window size for SAR image coregistration. Electronics Letters. 2003; **39**:122-124. DOI: 10.1049/el:20030054

[2] Zou W, Li Y, Li Z, Ding X. Improvement of the accuracy of InSAR image co-registration based on tie points: A review. Sensors. 2009;**9**: 1259-1281. DOI: 10.3390/s90201259

[3] Giles AB, Massom RA, Warner RC. A method for sub-pixel scale feature-tracking using Radarsat images applied to the Mertz Glacier Tongue, East Antarctica. Remote Sensing of Environment. 2009;**113**:1691-1699. DOI: 10.1016/j.rse.2009.03.015

[4] Li D, Zhang Y. A fast offset estimation approach for InSAR image subpixel registration. IEEE Geoscience and Remote Sensing Letters. 2012;**9**: 267-271. DOI: 10.1109/LGRS.2011. 2166752

[5] Harris C, Stephens M. A combined corner and edge detector. In: Proceedings of the Alvey Vision Conference (AVC'88); 31 August–2 September 1998; Manchester, U.K. pp. 147-151

[6] Bentoutou Y, Taleb N, Kpalma K, Ronsin J. An automatic image registration for applications in remote sensing. IEEE Transactions on Geoscience and Remote Sensing. 2005; **43**:2127-2137. DOI: 10.1109/TGRS. 2005.853187

[7] Lowe DG. Distinctive image features from scale-invariant keypoints. International Journal of Computer Vision. 2004;**60**:91-110. DOI: 10.1023/B: VISI.0000029664.99615.94

[8] Bay H, Ess A, Tuytelaars T, Van Gool L. Speeded-up robust features (SURF). Computer Vision and Image Understanding. 2008;**110**:346-359. DOI: 10.1016/j.cviu.2007.09.014

[9] Lindeberg T. Feature detection with automatic scale selection. Internatioal Journal of Computer Vision. 1998;**30**: 79-116. DOI: 10.1023/A:1008045108935

[10] Mikolajczyk K, Schmid C. A performance evaluation of local descriptors. IEEE Transactions on Pattern Analysis and Machine Intelligence. 2005;**27**:1615-1630. DOI: 10. 1109/TPAMI.2005.188

[11] Wessel B, Huber M, Roth A. Automatic image-to-image registration of near real-time SAR images. In: Proceedings of the Envisat Symposium; 23–27 April 2007; Montreux, Switzerland. pp. 1-6

[12] Li F, Zhang G, Yan J. Coregistration based on SIFT algorithm for synthetic aperture radar interferometry. In: Archives of International Society got Photogrammetry and Remote Sensing (ISPRS'08); 3–11 July 2008; Beijing, China. pp. 123-128

[13] Chen E, Li Z, Tian X, Li S. Application of scale invariant feature transformation to SAR imagery registration. Acta Automatica Sinica. 2008;**34**:861-868. DOI: 10.3724/SP. J.1004. 2008.00861

[14] Li D, Zhang Y. Geometric feature-based image coregistration approach for InSAR. In: Proceedings of the Asian-Pacific Conference on Synthetic Aperture Radar; 26–30 October, 2009; Xi'an, China. pp. 1026-1030

[15] Schwind P, Suri S, Reinartz P, Siebert A. Applicability of the SIFT operator to geometric SAR image registration. International Journal of

Remote Sensing. 2010;**31**:1959-1980. DOI: 10.1080/01431160902927622

[16] Wang S, You H, Fu K. BFSIFT: A novel method to find feature matches for SAR image registration. IEEE Geoscience and Remote Sensing Letters. 2012;**9**:649-653. DOI: 10.1109 / LGRS.2011.2177437

[17] Gong M, Zhao S, Jiao L, Tian D, Wang S. A novel coarse-to-fine scheme for automatic image registration based on SIFT and mutual Information. IEEE Transactions on Geoscience and Remote Sensing. 2014;**52**:4328-4338. DOI: 10.1109/TGRS.2013.2281391

[18] Dellinger F, Delon J, Gousseau Y, Michel J, Tupin F. SAR-SIFT: A SIFT-like algorithm for SAR images. IEEE Transactions on Geoscience and Remote Sensing. 2015;**53**:453-466. DOI: 10.1109/ TGRS.2014.2323552

[19] Wang F, You H, Fu X. Adapted anisotropic Gaussian SIFT matching strategy for SAR registration. IEEE Geoscience and Remote Sensing Letters. 2015;**12**:160-164. DOI: 10. 1109/LGRS. 2014.2330593

[20] Wang B, Zhang J, Lu L, Huang G, Zhao Z. A uniform SIFT-like algorithm for SAR image registration. IEEE Geoscience and Remote Sensing Letters. 2015;**12**:1426-1430. DOI: 10.1109/ LGRS.2015.2406336

[21] Zeng L, Zhou D, Liang J, Zhang K. Polar scale-invariant feature transform for synthetic aperture radar image registration. IEEE Geoscience and Remote Sensing Letters. 2017;**14**: 1101-1105. DOI: 10.1109/LGRS. 2017.2698450

[22] Ma W, Wen Z, Wu Y, Jiao L, Gong M, Zheng Y, et al. Remote sensing image registration with modified SIFT and enhanced feature matching. IEEE Geoscience and Remote Sensing Letters. 2017;**14**:3-7. DOI: 10.1109/LGRS. 2016.2600858

[23] Xiang Y, Wang F, You H. OS-SIFT: A robust SIFT-like algorithm for high-resolution optical-to-SAR image registration in suburban areas. IEEE Transactions on Geoscience and Remote Sensing. 2018;**56**:3078-3090. DOI: 10.1109/TGRS.2018.2790483

[24] Teke M, Temizel A. Multi-spectral satellite image registration using scale-restricted SURF. In: Proceedings of International Conference on Pattern Recognition (ICPR'10); 23–26 August 2010; Istanbul, Turkey. pp. 2310-2313

[25] Nguyen H-G, Fablet R, Ehrhold A, Boucher J-M. Keypoint-based analysis of sonar images: Applications to seabed recognition. IEEE Transactions on Geoscience and Remote Sensing. 2012; **50**:1171-1184. DOI: 10.1109/ TGRS.2011.2165848

[26] Liu R, Wang Y. SAR image matching based on speeded up robust feature. In: Proceedings of the WRI Global Congress on Intelligent System (GCIS'09); 19–21 May, 2009; Xiamen, China. pp. 518-522

[27] Lee WK, Kim AL. An efficient automatic geo-registration technique for high resolution spaceborne SAR image fusion. In: Proceedings of IEEE International Geoscience and Remote Sensing Symposium (IGASS'11); 24–29 July, 2011; Vancouver, Canada. pp. 3566-3569

[28] Zhang H, Ni W, Yan W, Wu J, Li S. Robust SAR image registration based on edge matching and refined coherent point drift. IEEE Geoscience and Remote Sensing Letters. 2015;**12**: 2115-2119. DOI: 10.1109/LGRS. 2015.2451396

[29] Li D, Zhang Y. On the appropriate feature for general SAR image registration. Proceedings of SPIE. 2012; **8536**:85360X. DOI: 10.1117/12.970520

[30] Fischler MA, Bolles RC. Random sample consensus: A paradigm for model fitting with applications to image analysis and automated cartography. Communications of the ACM. 1981;**24**: 381-395. DOI: 10.1145/358669.358692

[31] Zuliani M. Computational methods for automatic image registration [thesis]. Santa Barbara, U.S.: University of California at Santa Barbara; 2007

[32] Massart DL, Kaufman L, Rousseeuw PJ, Leroy A. Least median of squares: A robust method for outlier and model error detection in regression and calibration. Analytica Chimica Acta. 1986;**187**:171-179. DOI: 10.1016/S0003-2670(00)82910-4

[33] Rousseeuw PJ, Van Driessen K. Computing LTS regression for large data sets. Data Mining and Knowledge Discovery. 2006;**12**:29-45. DOI: 10.1007/s10618-005-0024-4

[34] Zhang Z, Deriche R, Faugeras O, Luong QT. A robust technique for matching two uncalibrated images through the recovery of the unknown epipolar geometry. Artificial Intelligence. 1995;**78**:87-119. DOI: 10.1016/0004-3702(95)00022-4

[35] Li D, Zhang Y. The appropriate parameter retrieval algorithm for feature-based SAR image registration. Proceedings of SPIE. 2012;**8536**:85360Y. DOI: 10.1117/12.970522

[36] Brown LG. A survey of image registration techniques. ACM Computing Surveys. 1992;**24**:325-376. DOI: 10.1145/146370.146374

[37] Zitova B, Flusser J. Image registration methods: A survey. Image and Vision Computing. 2003;**21**: 977-1000. DOI: 10.1016/S0262-8856(03)00137-9

[38] Dawn S, Saxena V, Sharma B. Remote sensing image registration techniques: A survey. In: Proceedings of International Conference on Image and Signal Processing (ICISP'10); 30 June–2 July, 2010; Trois-Rivieres, Canada. pp. 103-112

[39] Mikolajczyk K, Schmid C. An affine invariant interest point detector. In: Proceedings of European Conference on Computer Vision (ECCV'02); 28–31 May, 2002; Copenhagen, Denmark. pp. 128-142

[40] Rufino G, Moccia A, Esposito S. DEM generation by means of ERS tandem data. IEEE Transactions on Geoscience and Remote Sensing. 1998; **36**:1905-1912. DOI: 10.1109/36. 729362

[41] Li FK, Goldstein RM. Studies of multibaseline spaceborne interferometric synthetic aperture radars. IEEE Transactions on Geoscience and Remote Sensing. 1990; **28**:88-97. DOI: 10.1109/36.45749

[42] Brown M, Lowe DG. Invariant features from interest point groups. In: Proceedings of the British Machine Vision Conference (BMVC'02); 2–5 September, 2002; Cardiff, U.K. pp. 253-262

[43] Li D, Zhang Y. A novel approach for the registration of weak affine images. Pattern Recognition Letters. 2012;**33**: 1647-1655. DOI: 10.1016/j.patrec.2012.04.009

[44] Li Z, Bethel J. Image coregistration in SAR interferometry. In: Archives of International Society got Photogrammetry and Remote Sensing (ISPRS'08); 3–11 July 2008; Beijing, China. pp. 433-438

[45] Bamler R, Eineder M. Accuracy of differential shift estimation by

correlation and split-bandwidth interferometry for wideband and Delta-k SAR systems. IEEE Geoscience and Remote Sensing Letters. 2005;**2**:151-155. DOI: 10.1109/LGRS.2004.843203

[46] Lee JS, Pottier E. Polarimetric Radar Imaging: From Basics to Applications. Boca Raton, U.S.: CRC Press; 2009. pp. 143-177

[47] Li D, Zhang Y. A rigorous SAR epipolar geometry modeling and application to 3D target reconstruction. IEEE Selected Topics in Applied Earth Observations and Remote Sensing. 2013; **6**:2316-2323. DOI: 10.1109/JSTARS.2013.2249575

[48] Nitti DO, Hanssen RF, Refice A, Bovenga F, Nutricato R. Impact of DEM-assisted coregistration on high-resolution SAR interferometry. IEEE Transactions on Geoscience and Remote Sensing. 2011;**49**:1127-1143. DOI: 10.1109/TGRS.2010.2074204

[49] Gabriel AK, Goldstein RM. Crossed orbit interferometry: Theory and experimental results from SIR-B. International Journal of Remote Sensing. 1988;**9**:857-872. DOI: 10. 1080/01431168808954901

[50] Rousseeuw PJ, Leroy A. Robust Regression and Outlier Detection. Hoboken, U.S.: John Wiley & Sons, Inc.; 2005. pp. 1-157. DOI: 10.1002/0471725382

[51] Sweet AD, Dubbert DF, Doerry AW, Sloan GR, Dee Gutierrez V. A portfolio of fine resolution Ku-band MiniSAR images. In: Proceedings of SPIE Defense and Security Symposium; 17–21 April, 2006; Orlando, U.S. pp. 621002-621006

[52] Li D, Zhang Y. Unified Huynen phenomenological decomposition of radar targets and its classification applications. IEEE Transactions on Geoscience and Remote Sensing. 2016; **54**:723-743. DOI: 10.1109/TGRS.2015.2464113

[53] Li D, Zhang Y. Adaptive model-based classification of PolSAR data. IEEE Transactions on Geoscience and Remote Sensing. 2018;**56**:1-16. DOI: 10.1109/TGRS.2018.2845944

Chapter 3

L-Band SAR Disaster Monitoring for Harbor Facilities Using Interferometric Analysis

Ryo Natsuaki

Abstract

Synthetic aperture radar (SAR) has become a major tool for disaster monitoring. Its all-weather capability enables us to monitor the affected area soon after the event happens. Since the first launch of spaceborne SAR, its amplitude images have been widely used for disaster observations. Nowadays, an accurate orbit control and scheduled frequent observations enable us to perform interferometric analysis of SAR (InSAR) and the use of interferometric coherence. Especially for L-band SAR, its long-lasting temporal coherence is an advantage to perform precise interferometric coherence analysis. In addition, recent high resolution SAR images are found to be useful for observing relatively small targets, e.g., individual buildings and facilities. In this chapter, we present basic theory of SAR observation, interferometric coherence analysis for the disaster monitoring, and its examples for the harbor facilities. In the actual case, DInSAR measurement could measure the subsidence of the quay wall with 3 cm error.

Keywords: synthetic aperture radar (SAR), interferometry, interferometric coherence, disaster monitoring, infrastructure monitoring

1. Introduction

In the last decade, Interferometric Synthetic Aperture Radar (InSAR) has widely spread for measuring ground deformations caused by disasters, for example, earthquakes, volcanic eruptions, or subsidence [1, 2]. It can measure several centimeters of deformation with one pair of SAR images. The accuracy can be increased to several millimeters per year by applying time-series analysis [3, 4]. Compared with traditional optical or amplitude-based SAR analyses, e.g., [5], the advantages of InSAR-based monitoring are, for example, its sensitivity for the deformation and all-weather availability [6–8]. It is effective in the detection of various deformations caused by disasters such as earthquakes [9, 10], volcanic eruptions [11], storms [12], and human disaster [13]. Especially for the long wavelength SAR, i.e., L-band SAR, its long-lasting temporal coherence enables us to perform precise multitemporal interferometric coherence analysis [14]. Polarimetric analysis (PolSAR) has been also proposed for the damage detection using scattering mechanism analysis [15]. In PolSAR mode, SAR transmits both horizontal and vertical polarized waves and receives their co- and cross-polarized signals to see the scattering mechanism of targets. The collapsed buildings show different scattering mechanisms when they are compared with standing buildings. The most important examples are derived from 2011 off the pacific coast of Tohoku Earthquake [16–19]. Current problem for this method in the operational SARs is that there is less acquisition for full polarimetric mode, narrower swath width, and less spatial resolution caused by the operational limit of the platforms.

Another merit of spaceborne observations is that they have a wider observation swath than that of airborne observations, resulting in faster measurement over a wide area. However, the area of deformation is assumed to be larger than hundreds of meters in the world of InSAR. If its spatial resolution increases, it can be applied to smaller targets' deformation, e.g., the disaster monitoring of harbor facilities smaller than 100 m [20], in addition to the existing change detection methods, e.g., [21, 22].

Currently, operational SAR satellites aim wide swath or high resolution. ALOS-2, COSMO-SkyMed, RADARSAT-2, and TerraSAR/TanDEM-X aim higher resolution (<5 m) with relatively narrow swath width (<50 km), while Sentinel-1 aims wider swath width (200 km) with lower resolution (>20 m). Observation with higher resolution can achieve precise texture of the ground. One can analyze individual buildings with such a high resolution, while preceding researches mostly aim to evaluate in the size of a city block [23, 24]. On the other hand, wider observation swath is required for frequent and scheduled global observation using small number of satellites. That is, the higher frequent acquisition enables us to analyze the region of interest (RoI) without making any conflict with other observation requirements. The frequent observation is a requirement not only for time-series analysis but also for disaster monitoring that the users must observe the affected area as soon as possible. In the next decade, wide swath and high resolution are going to be combined, and the Earth will be observed weekly or bi-weekly with higher than 10 m resolution by SAR satellites such as ALOS-4, NISAR, Sentinel-1 NG, and TanDEM-L.

In such an era, disaster monitoring with SAR data using interferometric analysis becomes more useful [25]. In addition to the traditional amplitude-based change detection, centimeter-order deformation detection and interferometric coherence-based damage assessment will be more operational. One can acquire a delineation map over dozens of square kilometers for the affected area with a few meters resolution several hours after the observation, which cannot be achieved with ground/airborne surveys.

Monitoring harbor facilities plays an important role in the recovery phase in the disaster, because maritime traffic is a backbone of the logistics. For example, a heavy storm may have damaged the seawalls and piers. However, it is difficult to assess the stability of them soon after the event by humans because the ocean is still heavy. A catastrophic earthquake and tsunamis may have damaged a number of harbors simultaneously. In such a case, the authorities have to assess the damage of their facilities and decide whether to rearrange the route. SAR can quickly observe the affected area remotely on behalf of the risky direct observation by humans. This is the reason why SAR can play an important role in the rescue and recovery phase of the disaster. This chapter thinks of it.

In this chapter, we firstly present a fundamental theory for the interferometric analysis of SAR. It includes the basis of differential InSAR (DInSAR) and interferometric coherence analysis. Next, we describe a basic scheme of harbor monitoring for disaster monitoring. Finally, we show several examples in the real case, including the latest L-band SAR satellite Advanced Land Observing Satellite-2 (ALOS-2 or DAICHI-2) [26].

2. Synthetic aperture radar and interferometric analysis

In this section, we briefly explain the system of synthetic aperture radar (SAR) and its interferometric analysis (InSAR), including coherence analysis. Three monitoring methods are shown here. Firstly, wreckages and inundated area detection using amplitude information are described. Next, DInSAR-based displacement detection of the ground is explained. Finally, assessment for the damaged buildings from interferometric coherence is introduced. The descriptions are especially supposed for damage detection in harbor facilities, which is a combination of InSAR and coherence analysis.

2.1 Synthetic aperture radar

The larger antenna diameter derives the higher spatial resolution of radar systems. However, some platforms such as aircraft and satellites cannot deploy a sufficient size of the antenna because of their payload limitations. SAR solves this problem by moving itself and synthesizes the received signals by assuming that the ground targets are fixed [27].

If a SAR can use wide bandwidth, e.g., 1 GHz, it can achieve approximately 0.25 m of spatial resolution in the range direction. A typical high resolution SAR achieves 3–5 m. In the azimuth direction, a pulse repetition frequency (PRF) and aperture length are the large factors. The amplitude of a pixel of SAR image depends on the backscattering coefficient, and the phase depends on the distance between SAR and scatter. The phase information is difficult to handle because the wavelength is too short to measure the ground directly. On the other hand, the phase contains topographic, deformation, and the other valuable information. Those can be analyzed by interferometric analysis.

2.2 Differential interferometric SAR

Figure 1 presents a schematic diagram of InSAR analysis. A SAR image contains amplitude and phase information, in other words, complex-valued information and thus is called Single Look Complex (SLC) image. When we observe the same place from the same orbit multiply and multiply one SLC image (master) and another complex conjugated SLC image (slave), we can make an interferogram. The phase value of the interferogram, φ, is the phase difference between the master and slave. A SAR interferometric phase contains topographic, deformation, ionospheric delay, and tropospheric delay information [28].

In this chapter, we consider that the interferogram, φ, consists of the topographic, φ_{topo}, and deformation, φ_{defo}, components and ignore the others. The topographic component can be estimated by calculating the relationship between the known heights H acquired from a known topographic map as shown in Eq. (1).

$$\varphi_{topo} = \frac{4\pi B_{CT}\cos\left(\theta - \gamma_{CT}\right)}{\lambda R_m \sin\theta} H \tag{1}$$

where λ is the wavelength of SAR, $B_{CT}\cos(\theta\text{-}\gamma_{CT})$ is the perpendicular baseline of the two observations, θ is the incidence angle, and R_m is the slant range distance. Therefore, we can subtract φ_{topo} from the interferogram and measure the deformation component. If deformation occurs between two observations, the deformation phase value, φ_{defo}, corresponds to the shrink or extension in the line-of-sight

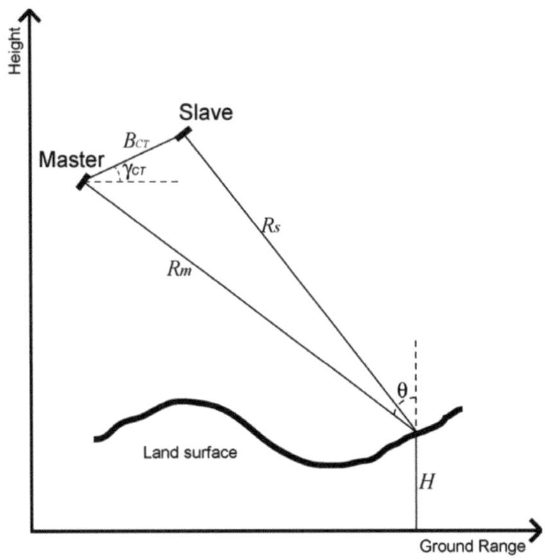

Figure 1.
Schematic diagram of InSAR analysis.

distance between the satellite and ground targets. If we denote the change in the line-of-sight distance as ΔR, the phase value can be calculated with the wave length λ using Eq. (2).

$$\varphi_{defo} = \frac{4\pi}{\lambda} \Delta R \tag{2}$$

As shown in the Eq. (2), the range of phase value is limited between $-\pi$ and π, and thus, we cannot distinguish deformations greater than a quarter wavelength. For example, L-band radar has approximately 24 cm wavelength. When $\Delta R = 0$, $+/-12$, $+/-24$... cm, φ_{defo} becomes 0 with indefinite $2n\pi$. Therefore, we cannot define the deformation if the neighboring pixels have more than a 6 cm line-of-sight difference. Long wavelength has an advantage to measure a large deformation. We can measure the absolute deformation by unwrapping the phase as long as the deformation satisfies the sampling theorem. The robustness of DInSAR analysis for harbor facilities is discussed in [20]. In [20], it is reported that the average error of the 11 observations was 0.1 cm, and its standard deviation was 0.4 cm for ideal case. That is, there was no systematic error of more than 0.1 cm when we apply the averaging filter when it contains a 0.4 cm variation inside the averaging window. At the same time, the average of the standard deviation of every observation was 1.0 cm. That is, a measured deformation with L-band SAR contains a 1.0 cm error.

2.3 Phase unwrapping

In order to calculate the absolute amount of the deformation, phase unwrapping process is required. The exact operation of phase unwrapping is a line integration of the phase values. We can achieve the integration result, which is independent of the integration path, as long as the amount of deformation between all neighboring pixels satisfies the sampling theorem. If there are residual points, i.e. rotational points, the unwrapping results become dependent on the integration path.

To solve this problem, estimating an appropriate phase value (filtering) and finding a specific integration path is required. Various filtering and unwrapping methods have been proposed. The Goldstein-Werner filter [29] is the famous low pass filter in the frequency domain. Probability estimation methods such as Markov random field model [30] and Bayesian estimation [31] have been proposed too. Nonlocal filter is widely used for its robustness [32]. Robust unwrapping methods have also been proposed. Branch-cut technique [33] tries to find the minimum cost to cancel the SPs by connecting opposite rotation side ones. Least square methods [34, 35] use Fourier transformation to distinguish steep slope from high frequency noise. The singularity spreading technique [36] is a newly developed method, which simply cancels residues by adding opposite direction to send residue to the other residues. In this chapter, we applied Markov random field model [30] filter and a least-square method [35] for phase unwrapping.

2.4 Interferometric coherence analysis

Interferometric coherence represents the uniformity of the interferometric pair of the SAR images [37, 38]. Interferometric coherence becomes high when the Master and Slave images are close to each other, while it decreases when two are completely different. The coherence value is calculated from the cross-correlation and autocorrelation between the two observations as shown in Eq. (3). When the ground targets are damaged or collapsed by disasters or human activities, the identical position reflects radio waves differently when we compare the pre- and post-event SLCs. In this case, the interferometric phase value contains no information, and the signal in master and slave SLC has no correlation. If the ground surface has been changed by the disaster, this effect appears as a large decrease in interferometric coherence.

$$\gamma = \frac{\langle M*\overline{S}\rangle}{\sqrt{\langle M*\overline{M}\rangle}\sqrt{\langle S*\overline{S}\rangle}}, \tag{3}$$

where M and S represent Master and Slave samples, $M*\overline{M}$ represents the complex conjugate multiplication of M, and $< >$ represents the ensemble average of the samples in $< >$. In short, γ is a normalized cross-correlation of M and S, and thus, it varies from 0 to 1. $\gamma = 1$ only happens when $M = S$ and $\gamma = 0$ never occurs because of randomness. A large facility has a relatively high (approximately 0.7–0.9) value. Contrarily, bare soil and concrete caissons have low (0.3<) values because they have smooth surface and low reflectivity in radar. An insufficient window size will overestimate the coherence value, while the larger window size will reduce the ground resolution. The window size of the ensemble average in Eq. (3) is 5×5 pixels in this chapter. The interferometric coherence largely depends on the surface roughness and temporal stability. If it is too smooth and/or unstable, such as water surface, concrete surface, and highly active region, the radio wave does not return to the satellite coherently.

Coherence γ also depends on the interval of M and S. Though it is stable, the ground surface changes time by time. If master and slave images are acquired in, for example, different years, γ becomes lower. This is called temporal decorrelation. To avoid this effect, it is required to observe the same place frequently. In general, a lower band SAR has a slower temporal decorrelation.

When we compare γ of two interferograms, we can detect the effect of disasters. This is called multitemporal coherence analysis; its aim is to detect the damaged part from the change in γ. If we have at least one interferogram prior to the disaster, a pre-event interferogram and an interferogram which is made from pre-event and post-event SLCs, a co-event interferogram, we can compare their coherence values.

This is called multitemporal interferometric coherence change detection. If the co-event interferometric coherence is lower than the pre-event one, though considering temporal decorrelation, it can be regarded that scatterers on the surface have been damaged and/or moved largely. Two definitions for the coherence decrease $d\gamma$ can be considerable. One is the simple difference or un-normalized coherence decrease (CD) $d\gamma_{un}$, and the other is the normalized coherence decrease $d\gamma_{norm}$, as shown in Eqs. (4) and (5).

$$d\gamma_{un} = \gamma_{pre} - \gamma_{co} \quad (4)$$

$$d\gamma_{norm} = \frac{\gamma_{pre} - \gamma_{co}}{\gamma_{pre} + \gamma_{co}} \quad (5)$$

where the pre-event coherence is γ_{pre} and the co-event coherence is γ_{co}. Eq. (4) requires relatively large γ_{pre} and cannot be applied for low coherency areas, such as vegetated ground. On the other hand, Eq. (5) does not require large γ_{pre}, while the temporal decorrelation will suffer in accuracy. In this chapter, Eq. (4) is applied because harbor facilities generally have large γ_{pre} in long temporal baseline.

The facilities should be regarded as damaged when $d\gamma_{un}$ exceeds the specific threshold. The threshold is generally defined manually to reduce the effect of temporal and spatial decorrelations. In [39], the authors found that the buildings which are larger than $200m^2$ can be evaluated by setting the threshold 0.3. When the buildings exceed the threshold, they were moderately or severely damaged when they are classified by EMS-98 scheme [40]. In this chapter, therefore, threshold for $d\gamma_{un}$ is set to 0.3.

2.5 Scattering mechanisms of harbor facilities

In order to apply those analyses above to the disaster monitoring of harbor facilities, the scattering mechanisms of SAR are briefly described. Numerical models are the same to the other cases; however, it is worth mentioning what the scatterers are in the harbor. **Figure 2** shows the schematic image of harbor. In the figure, SAR satellite is observing the harbor from left top of the figure.

Figure 2(a) shows the scattering mechanisms in normal condition. Region A is water, and therefore, backscattering coefficient is very low. Region B is the bare ground, and its brightness depends on its roughness. If the ground is covered by concrete or asphalt, it can be seen dark as same as water. Region C is layover of the facilities, and its surface scattering from the roof can be seen. On the other hand, Region D is a shadow region and Point Z cannot be observed by the satellite. Water, ground, and vertical walls work as smooth surface, and therefore, double-bounce effects are seen in Point X and Y.

Figure 3 presents an example of airborne L-band SAR image and the corresponding spaceborne optical image of the harbor. There are multiple bright targets which exist at the edge of the pier and the buildings. Those are the double-bounce effects. On the other hand, the top of the pier which is covered by asphalt and concrete is mostly dark as the water. The brightness of the rooftop of buildings depends on the structure of them. Those bright scatterers which do not move between two observations have high coherence. Note that ships have no coherence because they move on the water. Other moving facilities such as cars, containers, and cranes do not have high coherence too. On the other hand, stable facilities, e.g., buildings, walls, and vegetation have high coherence in L-band SAR. In short, interferometric analysis is applicable only to the stable bright scatterers.

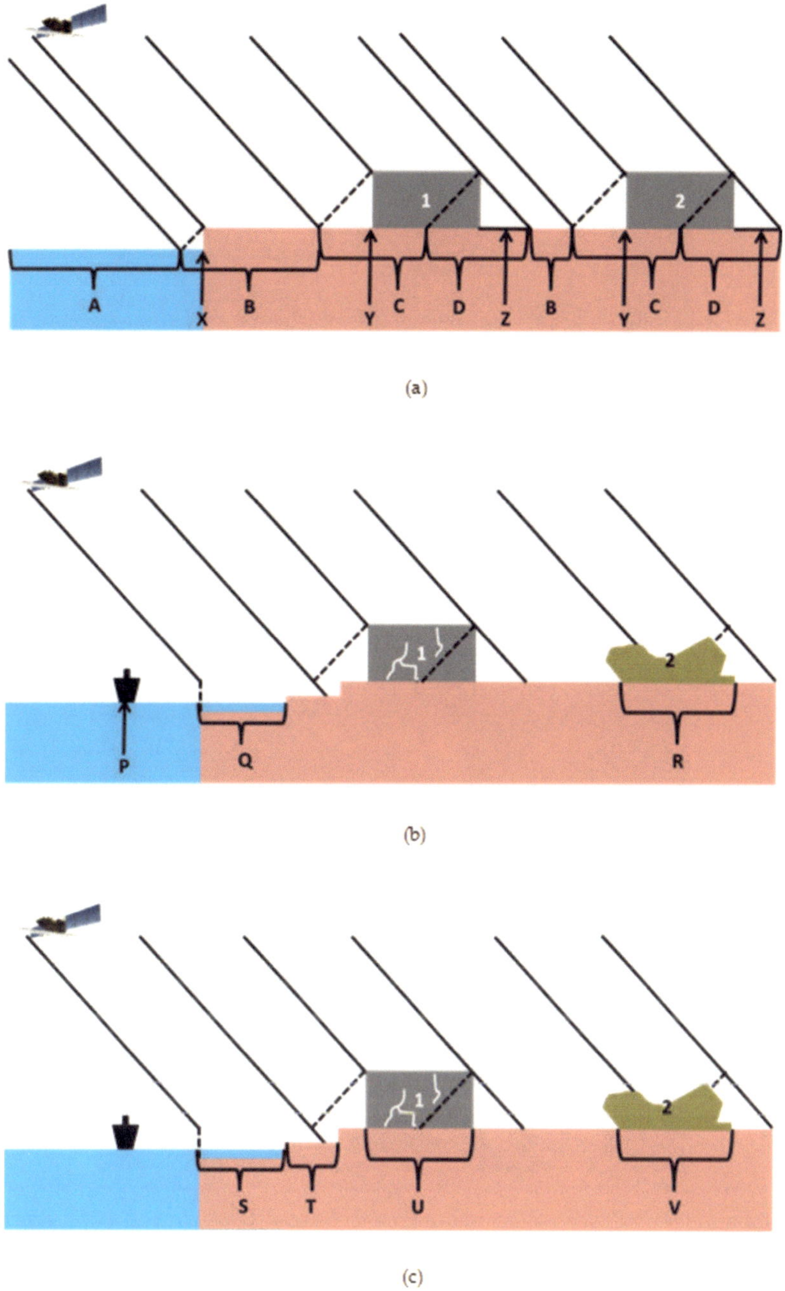

Figure 2.
Schematic image for disaster monitoring of harbor facilities using SAR. (a) Scattering mechanism in harbor and change detection based on (b) amplitude and (c) interferometric analysis.

Figure 2(b) shows a schematic diagram of the traditional amplitude-based SAR analysis. By comparing the pre- and post-disaster observation, the followings can be found. Object P wreckages on the water reflects the radar signal and appears in the post-disaster image. Region Q, an inundated area, will decrease the backscattering coefficients, and therefore, it appears as water region in the postdisaster image. Roads and other smooth surfaces are originally dark in SAR image and therefore may not change the amplitude by flooding. In addition, insufficient amount of subsidence or deformation will not be detected from amplitude image. Region R,

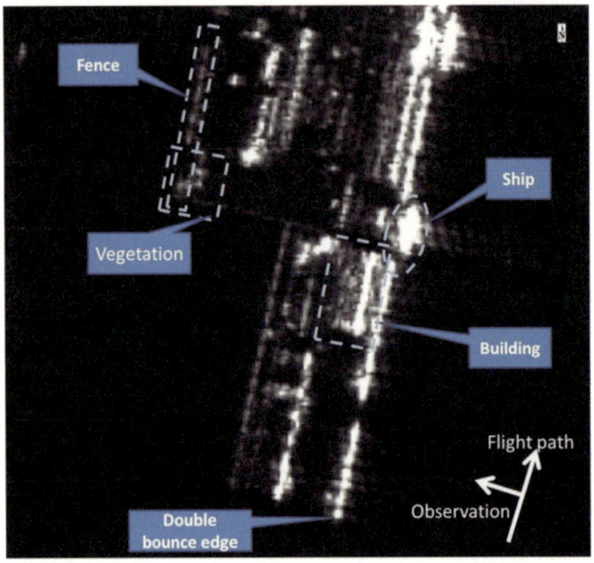

(a)

(b)

Figure 3.
Example of (a) SAR and (b) optical observation among harbor facilities.

the totally collapsed buildings, can be seen as a brighter scatterer in the postdisaster image, because in general, demolished buildings are more random surface than standing ones. On the other hand, slightly damaged buildings do not show any significant change in the amplitude image. In short, amplitude-based analysis can detect significant difference of the scatterers.

Figure 2(c) shows what can be observed by interferometric analysis. As water and surface has no coherency, wreckages on the water are not visible by interferometric analysis. Inundated Region S shows significant drop of coherence. Region T, deformation of the ground including subsidence or lateral flow, can be seen by interferometric phase and can be measured how large the surface moved. To calculate the absolute amount of the deformation, it requires relatively high coherence, that is, the place where the deformation can be measured has less damage. Region U and V, moderately damaged and collapsed buildings, can be detected by interferometric coherence analysis too. Its sensitivity is discussed precisely in [39].

In summary, some damages can be detected only by amplitude information, while the others can be detected only by interferometric analysis. Precise centimeter-order deformation can be measured when the surface keeps enough coherencies.

In a qualitative manner, one can segmentalize the disaster affected harbor facilities as following features.

- Nonaffected area can be recognized as high coherence and same amplitude areas.
- Deformation can be measured by DInSAR as long as the surface keeps enough coherence. Note that the phase component of the interferogram is relative value in the line-of-sight direction and not the absolute deformation of neither vertical nor horizontal direction.
- Moderately damaged buildings can be found by the decrease of coherence.
- Severely damaged buildings can be found by both decrease of coherence and increase of amplitude.
- Inundated areas appear as significant drop of amplitude as well as decrease of coherence.
- Wreckages on the water can be found from increase of the amplitude in the water region.

3. Damage detection scheme

Here, a brief detection scheme is introduced. In the rescue and recovery phase of disaster, mapping an affected area is one of the urgent tasks. The authorities use the delineation map for planning their activities. However, not all the responsible persons are familiar with remote sensing, especially for SAR. Therefore, intuitive classification is required.

Figure 4 shows an example of the classification flow. The classification scheme consists of five processes. First, the region which amplitude dropped more than 6 dB than pre-disaster data or weaker than the known water region is regarded as under the water and indicated as blue on map. The area which was affected by subsidence, tidal wave, and/or tsunami will be visualized.

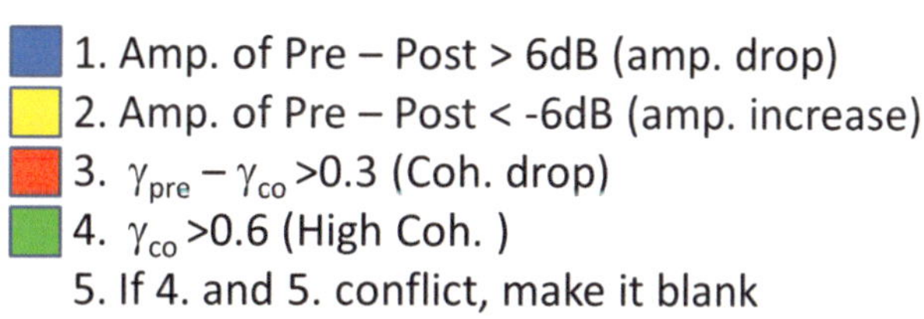

Figure 4.
Decision flow of the quick assessment.

Second, the region in which amplitude increased more than 6 dB than pre-disaster data is indicated as yellow on map. Wreckages and totally collapsed buildings will appear here.

Third, the region which coherence dropped more than 0.3 than pre-disaster dataset is regarded as inundated and indicated as red on map. Moderately or severely damaged facilities will be shown in this color. Totally collapsed buildings will be also classified here too.

Fourth, the region in which coherence is higher than 0.6 is regarded as not affected and indicated as green. Showing "safe zone" is demanded for the authorities to decide from where they start their operations.

If the region fulfills both 3 and 4, they do indicate none of them. Some large buildings have higher than 0.9 of coherence in pre-disaster pair, and their coherence may keep higher than 0.6 in co-disaster pair. In such a case, it is difficult to distinguish whether the buildings are damaged or not.

Measurement results for the deformation by DInSAR are presented in the different layer.

4. Examples

In this section, we applied the classification scheme to two examples. First is Ishinomaki port, Japan, which was severely affected by the 2011 off the Pacific coast of Tohoku earthquake. The port was observed by ALOS that was operated by JAXA until 2011. The second example is Kumamoto port, Japan, which was slightly affected by 2016 Kumamoto earthquake and observed by the latest ALOS-2. Note that the scheme can be applied to the other disasters such as typhoon and is evaluated in [20].

In Kumamoto port case, we measured the lateral flow with DInSAR. As the area is enough small, we assumed that all phase components consist of topography and deformation and ignored other phase components in the interferogram, such as the ionospheric [41–43] and tropospheric [44, 45] delays.

4.1 Ishinomaki port, Japan in 2011 off the Pacific coast of Tohoku earthquake

Ishinomaki port, Miyagi prefecture, Japan, was severely affected by the 2011 off the Pacific coast of Tohoku earthquake on March 11, 2011. A large tsunami hits the port, and almost all facilities were collapsed, damaged, or flushed out. The affected area was observed by ALOS several times. Here, we use the data set of Path 402, Frame 760. The observation dates are April 1, 2011, August 14, 2010, and May 14, 2010. The first two are used for co-event pair, and the latter two are for pre-event pair. Thanks to the L-band SAR's long lasting coherence, 8 months interval pair can be used effectively for the analysis. ALOS has approximately 10 m by 5 m resolution and therefore hardly investigate an identical building in general. However, the harbor facilities in this port were enough large to be distinguished. On the other hand, the deformation itself was too large to be measured by DInSAR. Therefore, we present a delineation map only.

Figure 5 presents the damage assessment results for the Ishinomaki port. As shown in the figure, most part of the coast line of the city was colored in red; the buildings are detected as moderately or severely damaged. In the left side of the figure, there is a large inundated area which is colored in blue. From **Figure 5**, it is also clear that some buildings on the hills in the north part of the city survived from the earthquake and tsunami.

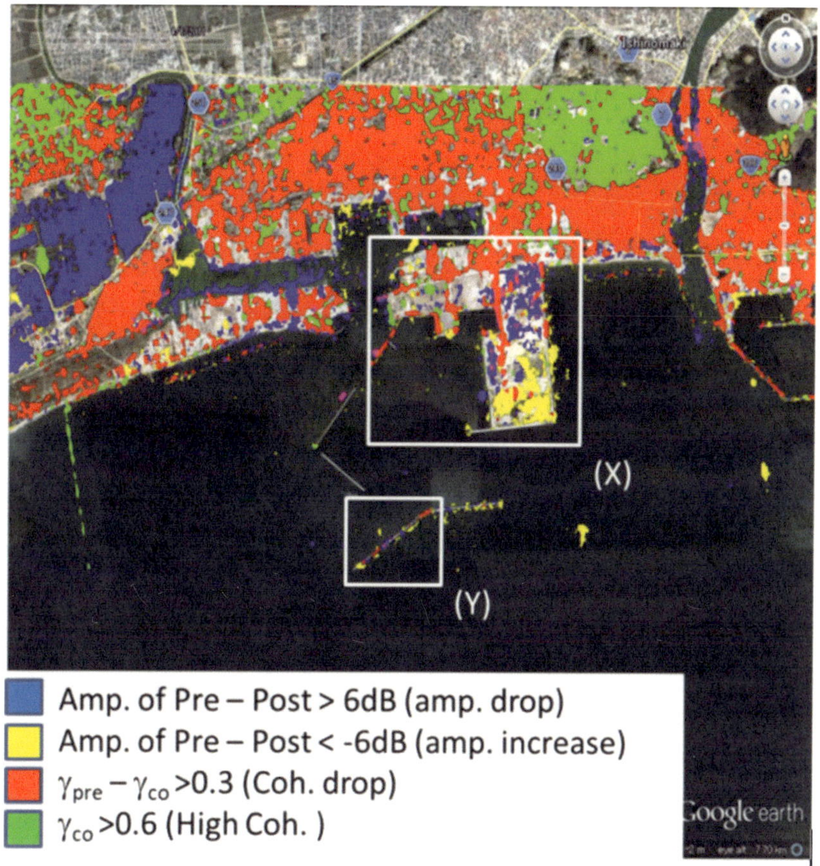

Figure 5.
Damage assessment results for Ishinomaki port, Japan, in 2011 off the Pacific coast of Tohoku earthquake.

Figure 6 shows the closed-up images with optical data of white rectangles, which are marked as X and Y in **Figure 5**. **Figure 6(a)** is a closed-up image for the piers and its comparison in optical images in Region X. Most facilities are demolished, while some of them are remaining. The pier in right hand side has both inundated and wreckage-covered area.

Figure 6(b) is a closed-up image and its optical comparison for the Region Y in **Figure 5**. This part is a breakwater of the port. Soon after the disaster, it is sometimes difficult to approach the offshore facilities. On the other hand, satellite-based SAR can observe them. In this case, tsunami hits the breakwater and some of them are sunk under the water. Wreckages are also found surrounding them. Most damaged buildings were found by interferometric coherence analysis. This is probably caused by the orientation, size, and structure of the buildings.

4.2 Kumamoto port, Japan in 2016 Kumamoto earthquake

Kumamoto port, Kumamoto prefecture, Japan, was hit by the earthquake on April 15, 2016. ALOS-2 had observed the port half a day before the earthquake and observed there again in the next revisit cycle (14-days) on April 29, 2016. There is another observation record from the same orbit on November 14, 2014, and therefore, we can perform the interferometric coherence analysis. The path and frame number of the observation are Path 28 and Frame 2930, respectively.

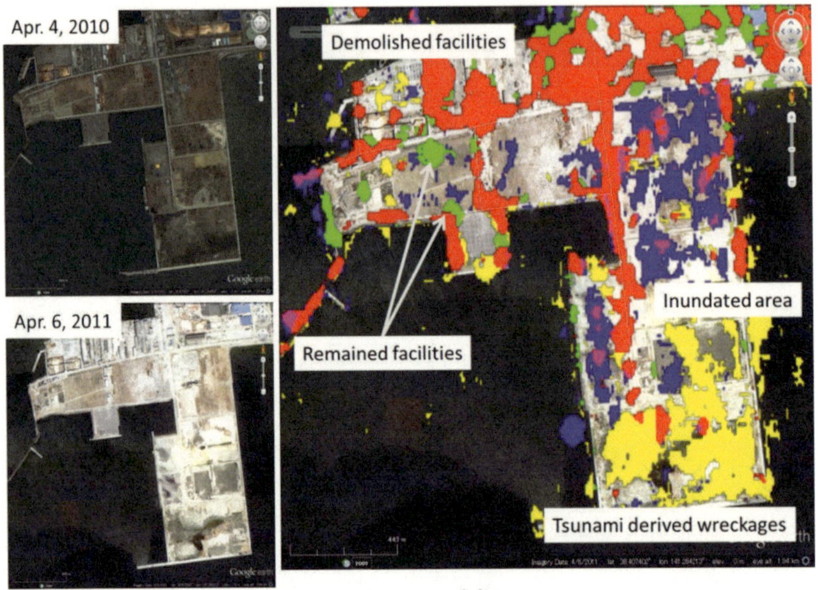

(a)

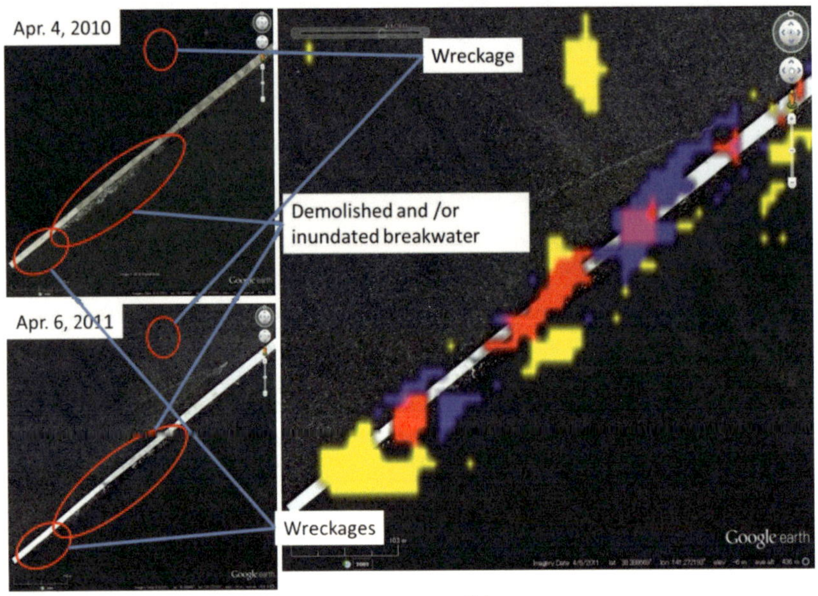

(b)

Figure 6.
Close up images for ***Figure*** *5: (a) Region X and (b) Region Y.*

Figure 7 shows an example of the interferometric phase and the position of the port. Each fringe of the interferometric phase represents 12 cm deformation. Fortunately, the port is enough far from the epicenter. Its overall deformation was small enough to continue the operation. **Figure 8** presents the delineation map and the analytical results for coherence analysis and DInSAR measurement. **Figure 8(a)** shows the delineation map. Fortunately, most part of the pier received no damage. Therefore, we could only detect the difference of the

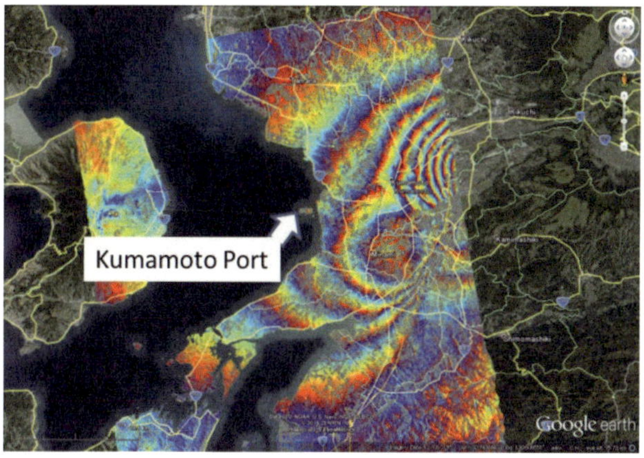

Figure 7.
Interferogram in Kumamoto earthquake.

berthing ships between April 15 and 29. There were almost no damaged facilities, and therefore, it is hardly visible to see nongreen part. In **Figure 8(b)**, we show the coherence drop data, red color in **Figure 8(a)**. Now, it is visible that the right hand side of the pier has several damaged facilities. According to the rapid report from the port [46], some roads and facilities received several damages.

In this case, temporal baseline of γ_{pre} is almost 17 months. In such a long interval, a SAR, which uses higher frequency (e. g., X- or C-band), cannot achieve enough coherence to compare with γ_{co}. These results indicate that L-band SAR may observe the earth from additional incidence angle once in a several years in order to prepare for the disaster. If we have multiple archives from multiple incidence angles, the operator can mauve the satellite to observe the affected area as soon as possible and compare the observation results with the archives. Such operation will greatly help the corresponding authorities because they need not to wait for the next "scheduled" observation.

Kumamoto port had been under construction to landfill. A lateral flow occurred in the north part of the pier. **Figure 8(c)** shows the measurement results of the DInSAR. The unwrapped result shows more than 20 cm of line-of-sight displacement. As ALOS-2 observed the port from west of the port, west half of the port moved toward the satellite and the east half moved away from the satellite. On the other hand, the existing parts show only small deformations. For example, the quay wall showed 5–10 cm subsidence by DInSAR measurement. On the other hand, the actual measurement in [46] was 7 cm. Therefore, the error in this case was 3 cm, which is larger than the ideal case in [20] (1 cm). This is caused by, for example, filtering errors, unwrapping errors, or the randomness in the subsidence. In summary, DInSAR could measure the subsidence of the quay wall with several centimeter of error. The measured deformation can be used in the recovery phase of the disaster. As single DInSAR pair can measure the line-of-sight displacement, three dimensional measurements require more than three observations. Especially for the satellite SAR, it is difficult to measure north-south deformation by interferometric analysis because it usually flights polar orbit and can only observe from east or west. If the deformation is enough large to be detected by co-registration, pixel-offset method can be applied.

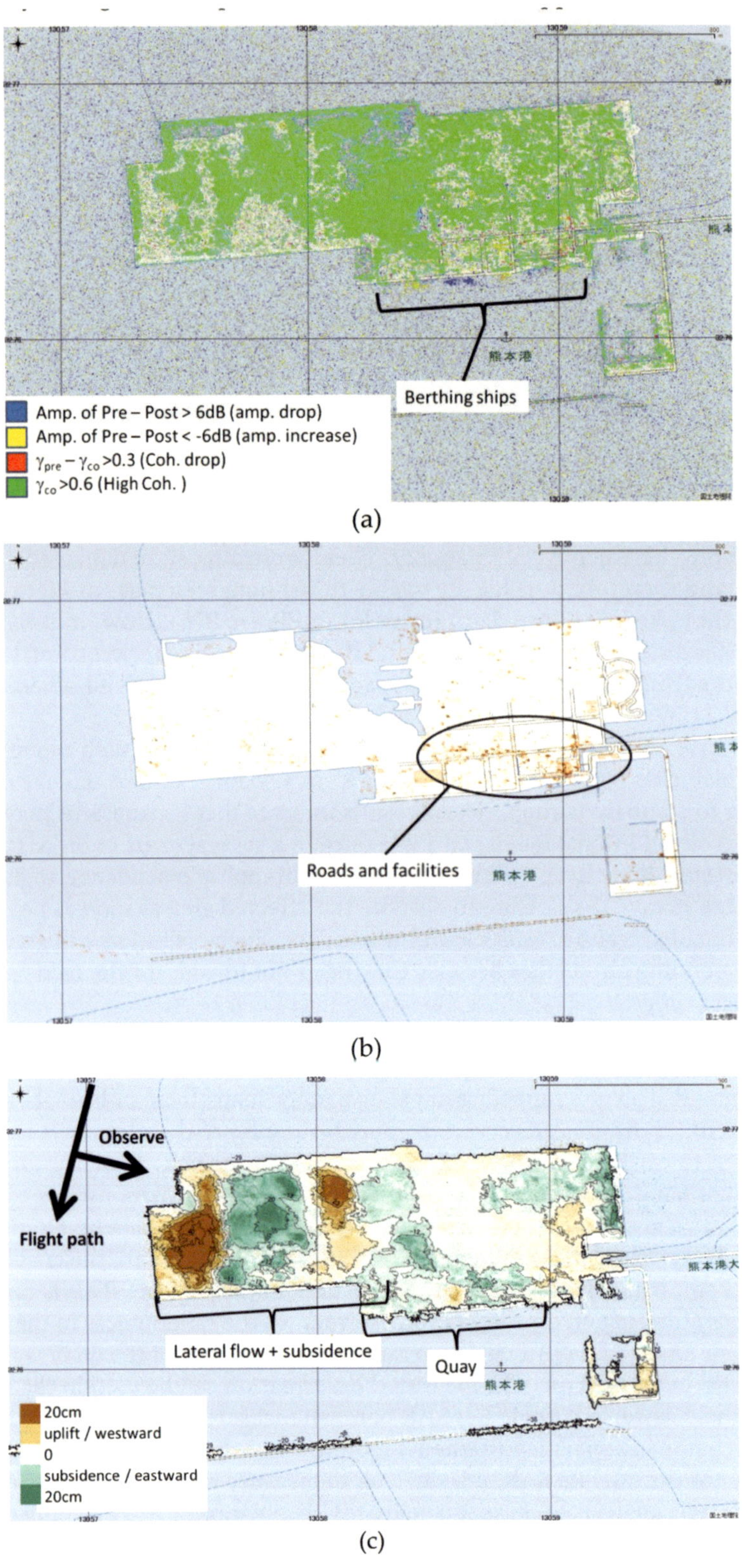

Figure 8.
Damage assessment results for Kumamoto port, Japan, in 2016 Kumamoto earthquake. (a) Damage assessment map, (b) coherence dropped part only, and (c) deformation map.

5. Conclusion

In this chapter, a rapid damage assessment scheme which is based on SAR interferometric analysis was introduced. With a combination of amplitude analysis, it is able to show an easy-understanding and enough-accurate delineation map. Furthermore, interferometric analysis can provide centimeter-order deformation map. In the example case of Kumamoto earthquake, ALOS-2 detected 5–10 cm subsidence in the quay wall, which was 7 cm in real measurement. These results are highly appreciated by the disaster corresponding authorities. In this chapter, a basic theory is shown. Its accuracy can be easily improved by, for example, machine learning and data-fusions with the other observations.

Acknowledgements

The ALOS/ALOS-2 original data are copyrighted by MEXT and JAXA, and provided under JAXA 6th ALOS Research Announcement PI No. 3044. The experiments were partially supported by the Council for Science, Technology, and Innovation, "Cross-ministerial Strategic Innovation Promotion Program (SIP), Infrastructure Maintenance, Renovation, and Management" (funding agency: NEDO). The maps in this chapter are provided from Geospatial Information Authority under the Ministry of Land, Infrastructure, Transport and Tourism of Japan. Google Inc. provided the aerial images on Google Earth. Dr. Natsuaki is currently with German Aerospace Center as a visiting scientist sponsored by JSPS Overseas Research Fellowships.

Conflict of interest

The author declares no conflict of interest.

Author details

Ryo Natsuaki
Department of Electrical Engineering and Information Systems, The University of Tokyo, Tokyo, Japan

*Address all correspondence to: natsuaki@eis.t.u-tokyo.ac.jp

References

[1] Boerner WM. Recent advances in extra-wide-band polarimetry, interferometry and polarimetric interferometry in synthetic aperture remote sensing and its applications, radar, sonar navigation. IEE Proceedings. 2003;**150**(3):113-124

[2] Plank S. Rapid damage assessment by means of multi-temporal SAR - a comprehensive review and outlook to Sentinel-1. Remote Sensing. 2014;**6**(6): 4870-4906. DOI: 10.3390/rs6064870

[3] Ferretti A, Prati C, Rocca F. Permanent sacatterers in SAR interferometry. IEEE Transactions on Geoscience and Remote Sensing. 2001; **39**(1):8-20

[4] Ferretti A, Fumagalli A, Novali F, Prati C, Rocca F, Rucci A. A new algorithm for processing interferometric data-stacks: SqueeSAR. IEEE Transactions on Geoscience and Remote Sensing. 2011;**49**(9):3460-3470

[5] Liu W, Yamazaki F, Adriano B, Mas E, Koshimura S. Development of building height data in Peru from high-resolution SAR imagery. Journal of Disaster Research. 2014;**9**(6): 1042-1049. DOI: 10.20965/jdr.2014. p1042

[6] Matsuoka M, Yamazaki F. Building damage mapping of the 2003 Bam, Iran, earthquake using ENVISAT/ASAR intensity imagery. Earthquake Spectra. 2005;**21**(S1):S285-S294

[7] Fielding EJ. Surface ruptures and building damage of the 2003 Bam, Iran, earthquake mapped by satellite synthetic aperture radar interferometric correlation. Journal of Geophysical Research. 2005;**110**(B03302)

[8] Arciniegas GA, Bijker W, Kerle N, Tolpekin VA. Coherence- and amplitude-based analysis of seismogenic damage in Bam, Iran, using ENVISAT ASAR data. IEEE Transactions on Geoscience and Remote Sensing. 2007; **45**(6):1571-1581

[9] Natsuaki R, Nagai H, Motohka T, Ohki M, Watanabe M, Thapa RB, et al. SAR interferometry using ALOS-2 PALSAR-2 data for the Mw 7.8 Gorkha Nepal earthquake, Earth, Planets and Space. 2016;**68-15**:1-13. DOI: 10.1186/ s40623-016-0394-4

[10] Jimenez C, Moggiano N, Mas E, Adriano B, Koshimura S, Fujii Y, et al. Seismic source of 1746 Callao earthquake from tsunami numerical Modeling. Journal of Disaster Research. 2013;**8**(2):266-273. DOI: 10.20965/ jdr.2013.p0266

[11] Hooper A, Sigmundsson F, Prata F. Remote sensing of volcanic hazards and their precursors. Proceedings of the IEEE. 2012;**100**(10):2908-2930

[12] Nakmuenwai P, Yamazaki F, Liu W. Multi-temporal correlation method for damage assessment of buildings from high-resolution SAR images of the 2013 typhoon Haiyan. Journal of Disaster Research. 2016;**11**(3):577-592. DOI: 10.20965/jdr.2016.p0577

[13] Liu W, Yamazaki F, Sasagawa T. Monitoring of the recovery process of the Fukushima Daiichi nuclear power plant from VHR SAR images. Journal of Disaster Research. 2016;**11**(2):236-245. DOI: 10.20965/jdr.2016.p0236

[14] Wei M, Sandwell DT. Decorrelation of L-band and C-band interferometry over vegetated areas in California. IEEE Transactions on Geoscience and Remote Sensing. 2010;**48**(7):2942-2952

[15] Yamaguchi Y. Disaster monitoring by fully polarimetric SAR data acquired with ALOS-PALSAR. Proceedings of the IEEE. 2012;**100**(10):2851-2860

[16] Sato M, Chen SW, Satake M. Polarimetric SAR analysis of tsunami damage following the march 11, 2011 East Japan earthquake. Proceedings of the IEEE. 2012;**100**(10):2861-2875

[17] Watanabe M, Motohka T, Miyagi Y, Yonezawa C, Shimada M. Analysis of urban areas affected by the 2011 off the pacific coast of Tohoku earthquake and tsunami with L-band SAR full-polarimetric mode. IEEE Geoscience and Remote Sensing Letters. 2012;**9**(3): 472-476

[18] Chen SW, Sato M. Tsunami damage investigation of built-up areas using multitemporal spaceborne full polarimetric SAR images. IEEE Transactions on Geoscience and Remote Sensing. 2013;**51**(4):1985-1997

[19] Chen SW, Wang XS, Sato M. Urban damage level mapping based on scattering mechanism investigation using fully polarimetric SAR data for the 3.11 East Japan earthquake. IEEE Transactions on Geoscience and Remote Sensing. 2016;**54**(12): 6919-6929

[20] Natsuaki R, Anahara T, Kotoura T, Iwatsuka Y, Tomii N, Katayama H, et al. Synthetic aperture radar interferometry for disaster monitoring of harbor facilities. Journal of Disaster Research. 2017;**12**(3):526-535

[21] Matsuoka M, Estrada M. Development of earthquake-induced building damage estimation model based on ALOS/PALSAR observing the 2007 Peru earthquake. Journal of Disaster Research. 2013;**8**(2):346-355. DOI: 10.20965/jdr.2013.p0346

[22] Adriano B, Mas E, Koshimura S, Estrada M, Jimenez C. Scenarios of earthquake and tsunami damage probability in Callao region, Peru using tsunami fragility functions. Journal of Disaster Research. 2014;**9**(6):968-975. DOI: 10.20965/jdr.2014.p0968

[23] Yonezawa C, Takeuchi S. Decorrelation of SAR data by urban damages caused by the 1995 Hyogokennanbu earthquake. International Journal of Remote Sensing. 2001;**22**:1585-1600

[24] Matsuoka M, Nojima N. Building damage estimation by integration of seismic intensity information and satellite L-band SAR imagery. Remote Sensing. 2010;**2**(9): 2111-2126

[25] Milillo P, Riel B, Minchew B, Yun SH, Simons M, Lundgren P. On the synergistic use of SAR constellations' data exploitation for earth science and natural hazard response. IEEE Journal of Selected Topics in Applied Earth Observations and Remote Sensing. 2016;**9**(3):1095-1100

[26] Arikawa Y, Saruwatari H, Hatooka Y, Suzuki S. PALSAR-2 launch and early orbit operation result. International Geoscience and Remote Sensing Symposium (IGARSS). 2014;**2014**: 3406-3409

[27] Cumming IG, Wong FHC. Digital Processing of Synthetic Aperture Radar Data: Algorithms and Implementation. Norwood, MA, USA: Artech House; 2005

[28] Ghiglia DC, Pritt MD. Two-Dimensional Phase Unwrapping: Theory, Algorithms, and Software. John Wiley & Sons, Inc.; 1998:31-58

[29] Goldstein RM, Werner CM. Radar interferogram filtering for geophysical applications. Geophysical Research Letters. 1998;**25**(21):4035-4038

[30] Yamaki R, Hirose A. Singular unit restoration in interferograms based on complex-valued Markov random field model for phase unwrapping. IEEE Geoscience and Remote Sensing Letters. 2009;**6**(1):18-22

[31] Ferraiuolo G, Poggi G. A Bayesian filtering technique for SAR interferometric phase fields. IEEE Transactions on Image Processing. 2004;**13**(10):1368-1378

[32] Sica F, Cozzolino D, Zhu XX, Verdoliva L, Poggi G. InSAR-BM3D: A nonlocal filter for SAR Interferometric phase restoration. IEEE Transactions on Geoscience and Remote Sensing. 2018; **56**:3456-3467

[33] Costantini M. A novel phase unwrapping method based on network programming. IEEE Transactions on Geoscience and Remote Sensing. 1998; **36**(3):813-821

[34] Pritt MD, Shipman JS. Least-squares two-dimensional phase unwrapping using FFT's. IEEE Transactions on Geoscience and Remote Sensing. 1994; **32**(3):706-708

[35] Suksmono AB, Hirose A. Progressive transform-based phase unwrapping utilizing a recursive structure. IEICE Transactions on Communications. 2006;**E89-B**(3): 929-936

[36] Yamaki R, Hirose A. Singularity-spreading phase unwrapping. IEEE Transactions on Geoscience and Remote Sensing. 2007;**45**(10):3240-3251

[37] Touzi R, Lopes A, Bruniquel J, Vachon PW. Coherence estimation for SAR imagery. IEEE Transactions on Geoscience and Remote Sensing. 1999; **37**(1):135-149. DOI: 10.1109/36.739146

[38] Abdelfattah R, Nicolas JM. Interferometric SAR coherence magnitude estimation using second kind statistics. IEEE Transactions on Geoscience and Remote Sensing. 2003; **44**(2):1942-1953

[39] Natsuaki R, Nagai H, Tomii N, Tadono T. Sensitivity and limitation in damage detection for individual buildings using InSAR coherence - a case study in 2016 Kumamoto earthquakes. Remote Sensing. 2018; **10**(2):245

[40] Grunthal G, editor. European Macroseismic Scale 1998. Luxembourg: Centre Europeen de Geodynamique et de Seismologie; 1998

[41] Mayer F, Bamler N, Jakowski R, Fritz T. The potential of low-frequency SAR systems for mapping ionospheric TEC distributions. IEEE Geoscience and Remote Sensing Letters. 2006;**3**(4): 560-564

[42] Gomba G, Parizzi A, De Zan F, Eineder M, Bamler R. Toward operational compensation of ionospheric effects in SAR interferograms: The split-spectrum method. IEEE Transactions on Geoscience and Remote Sensing. 2015; **54**(3):1446-1461. DOI: 10.1109/ TGRS.2015.2481079

[43] Jung HS, Lee WJ. An improvement of ionospheric phase correction by multiple-aperture interferometry. IEEE Transactions on Geoscience and Remote Sensing. 2015;**53**(9):4952-4960

[44] Doin MP, Lasserre C, Peltzer G, Cavalié O, Doubre C. Corrections of stratified tropospheric delays in SAR interferometry: Validation with global atmospheric models. Journal of Applied Geophysics. 2009;**69**:35-50. DOI: 10.1016/j.jappgeo.2009.03.010

[45] Bekaert DPS, Walters RJ, Wright TJ, Hooper AJ, Parker DJ. Statistical comparison of InSAR tropospheric correction techniques. Remote Sensing of Environment. 2015;**170**:40-47. DOI: 10.1016/j.rse.2015.08.035

[46] Nodu A. Damage Reports of Ports and Airports, Rapid Communication for 2016 Kumamoto Earthquake, Earthquake Engineering Committee, Japan Society of Civil Engineering. Available from http://committees.jsce.or.jp/eec2/node/76 [in Japanese, Accessed: 2018-08-31]

Section 2

Advanced Image Data Processing

Chapter 4

Utilization of Deep Convolutional Neural Networks for Remote Sensing Scenes Classification

Chang Luo, Hanqiao Huang, Yong Wang and Shiqiang Wang

Abstract

Deep convolutional neural networks (CNNs) have been widely used to obtain high-level representation in various computer vision tasks. However, for the task of remote scene classification, there are no sufficient images to train a very deep CNN from scratch. Instead, transferring successful pre-trained deep CNNs to remote sensing tasks provides an effective solution. Firstly, from the viewpoint of generalization power, we try to find whether deep CNNs need to be deep when applied for remote scene classification. Then, the pre-trained deep CNNs with fixed parameters are transferred for remote scene classification, which solve the problem of time-consuming and parameters over-fitting at the same time. With five well-known pre-trained deep CNNs, experimental results on three independent remote sensing datasets demonstrate that transferred deep CNNs can achieve state-of-the-art results in unsupervised setting. This chapter also provides baseline for applying deep CNNs to other remote sensing tasks.

Keywords: convolutional neural network, remote sensing, scene classification, deep learning, generalization power

1. Introduction

Remote sensing image processing achieves great advances in recent years, from low-level tasks, such as segmentation, to high-level ones, such as classification. [1–7] However, the task becomes incrementally more difficult as the level of abstraction increases, going from pixels, to objects, and then scenes. Classifying remote scenes according to a set of semantic categories is a very challenging problem, because of high intra-class variability and low interclass distance. [5–9] Therefore, the more representative and higher-level representations are desirable and will certainly play a dominant role in scene-level tasks. The deep convolutional neural network (CNN), which is acknowledged as the most successful and widely used deep learning model, attempts to learn high-level features corresponding to high level of abstraction [10]. Its recent impressive results for classification and detection tasks bring dramatic improvements beyond the state-of-the-art records on a number of benchmarks [11–14]. In theory, considering the subtle differences among categories in remote scene classification, we may attempt to form high-level representations for remote scenes from CNN activations. However, the acquisition of large-scale well-annotated remote sensing datasets is costly, and it is easy to over-fit when we try to train a high-powered deep CNN with small datasets in practice [15]. In other words, with limited remote sensing dataset, deep CNNs work perfectly on the training data but do not generalize well to test data, resulting in poor performance eventually.

ImageNet[1] is a large-scale dataset, which offers a very comprehensive database of more than 1.2 million categorized natural images of 1000+ classes [16]. Deep CNN models trained upon this dataset serve as the backbone for many segmentation, detection, and classification tasks on other datasets. Moreover, some very recent works have demonstrated that the representations learned from deep CNNs pre-trained on large datasets such as ImageNet can be transferable to image classification task [17]. Some works also start to apply them to remote sensing field and obtain state-of-the-art results for some specific datasets [15, 18, 19]. However, the generalization power of features learned from deep CNNs fades evidently when the features of remote sensing images become different with that of natural images in the ImageNet dataset [15, 18]. Therefore, to solve the problem discussed above, the generalization power of deep CNNs plays the key role. We find that the generalization power of a deep CNN is relative to its depth. A deeper architecture trained by large-scale dataset may lead to a more general hypothesis for remote scenes. To our surprise, features learned from deeper layers are more general than that learned from shallower layers in a deep CNN when we transfer them for remote scene classification. This overturns the traditional view that features in shallow layers of a deep CNN are composed of basic visual patterns (e.g., salient edges and borders) and they are more general for test data. Inspired by this, we evaluate the generalization power of transferred deep CNN for remote scenes in different conditions and explore the proper way to apply deep CNNs to remote scene classification with limited remote sensing data.

We conduct extensive experiments with transferred deep CNN and evaluate the generalization power of it on different remote sensing datasets that vary in space information. The results show that the depth of CNNs contributes to the generalization power of them. Features from deeper layers are more general for test data and brings better performance in remote scene classification. Then, we conduct extensive experiments with different pre-trained deep CNNs such as CaffeNet [13], GoogLeNet [20], and ResNet [21]. This chapter hardly contains any deep or new techniques, and our study so far is mainly empirical. However, a thorough report on generalization power of deep CNNs for remote scene classification has tremendous value for applying deep CNNs to remote sensing images. A satisfied answer to this question would not only help to make features of remote scenes more interpretable in deep CNNs, but it might also lead to more principled and reliable deep architecture design. Our main contributions are summarized as follows:

1. We thoroughly investigate how transferred deep CNNs work for remote scene classification with limited remote sensing data and how the generalization power of them affect their performance.

2. This chapter challenges the classical view of features learned in deep CNNs by showing that high-level features learned in deeper layers are more general than basic features (e.g., salient edges and borders) learned in shallower layers. Features learned in shallow layers of deep CNNs are not general enough for remote scenes. This leads us to believe that depth of CNNs enhances the

[1] http://www.image-net.org/challenges/LSVRC/

generalization power of the learned features and it is essential for remote scene classification.

3. Based on various pre-trained deep CNNs, we evaluate our proposed method on different remote sensing datasets that vary in space and spectrum. The results show that our proposed method can learn better features for remote scenes. With "unsupervised settings," our proposed method achieves state-of-the-art performance on some public remote scene datasets.

The rest of the chapter is organized as follows. Section 2 presents successful pre-trained deep CNNs nowadays and the way to transfer them for remote scene classification. Section 3 analyzes the generalization power of features in different layers of transferred deep CNN. Experiments are presented in Section 4, and we conclude the chapter in Section 5.

2. Transferred deep CNNs for remote scene classification

Convolutional neural networks are generally presented as systems of interconnected processing units which can compute values from inputs leading to an output that may be used on further units. The typical architecture of a deep CNN is composed of multiple cascaded layers with various types.

Among the different types of layers, the convolutional one is the responsible for capturing the features from the images. The first layers usually obtain low-level features (like edges, lines, and corners), while the others get high-level features (like structures, objects, and shapes). The process made in this type of layer can be decomposed into two phases: (i) the convolution step, where a fixed-size window runs over the image, with some stride, defining a region of interest and (ii) the processing step that uses the pixels inside each window as input for the neurons that, finally, perform the feature extraction from the region. The continuous form and discrete form of convolutional operation can be expressed as Eqs. (1) and (2), respectively:

$$h(x,y) = i*k(x,y) = \int_{-\infty}^{\infty}\int_{-\infty}^{\infty} i(u,v)k(x-u,y-v)dudv \tag{1}$$

$$H(x,y) = I*K(x,y) = \sum_m \sum_n I(m,n)K(x-m,y-n) \tag{2}$$

As to the input map, the convolutional operation can be further illustrated by **Figure 1**:

Conventionally, a nonlinear function is provided after the convolutional operation, which is usually called activation function. There are a lot of alternatives for activation function, such as sigmoid function $\frac{1}{1+e^{-x}}$ and tanh function $\frac{e^x-e^{-x}}{e^x+e^{-x}}$. The most popular activation function nowadays is called rectified linear unit (ReLU). ReLU has several advantages when compared to others: (i) works better to avoid saturation during the learning process, (ii) induces the sparsity in the hidden units, and (iii) does not face gradient vanishing problem as with sigmoid and tanh function. The mathematic form of the ReLU can be shown as follows:

$$a = \begin{cases} z, & if\ z>0 \\ 0, & otherwise \end{cases} \quad \Leftrightarrow \quad a = f(z) = \max(0,z) \tag{3}$$

Typically, after obtaining the convolved feature activations, we would next like to aggregate statistics of these features at various locations, and this aggregation

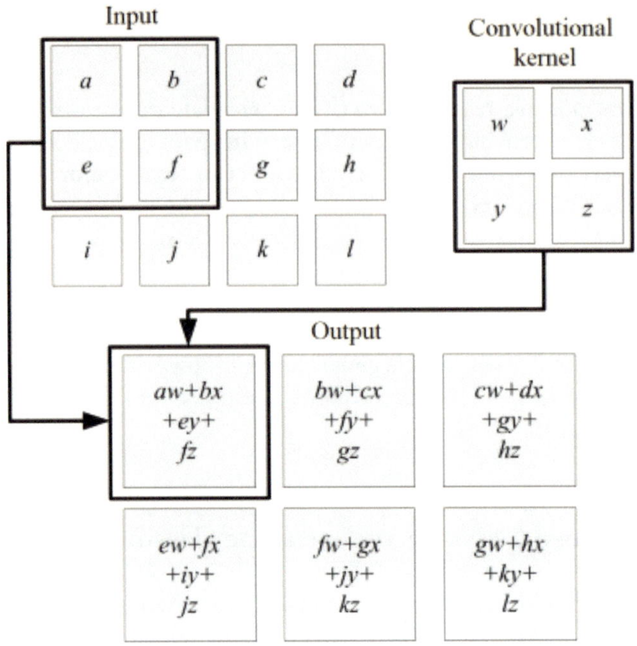

Figure 1.
Convolutional operation.

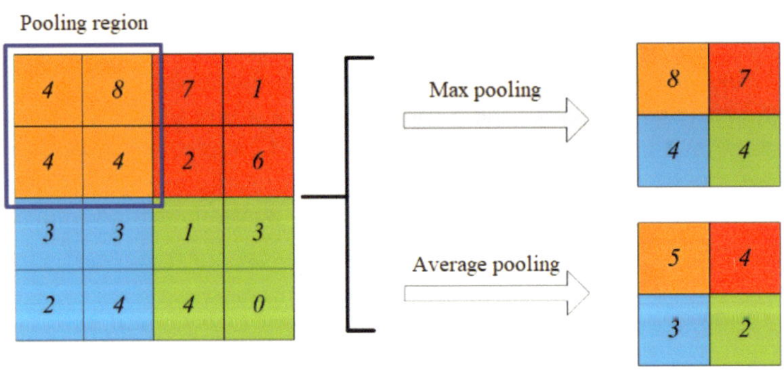

Figure 2.
Pooling operation.

operation is called pooling operation. Pooling operation within the pooling region translates convolved feature activations into pooled features, which are much lower in dimension and can improve classification results (i.e., less over-fitting). Pooling regions are usually contiguous areas in the convolved feature maps, and the pooled features are usually generated from the same filter. Then these pooled features would be "translation invariant." Although several novel pooling approaches have been proposed, max pooling and average pooling are still the most commonly used approaches as shown in **Figure 2**.

After several convolutional and pooling layers, there are the fully connected ones, which take all neurons in the previous layer and connect them to every single neuron in its layer. Since a fully connected layer occupies most of the parameters, over-fitting can easily happen. To prevent this, the dropout method was employed as shown in **Figure 3**. This technique randomly drops several neuron outputs, which do not contribute to the forward pass and backpropagation anymore. This neuron drops are equivalent to decreasing the number of neurons of the network,

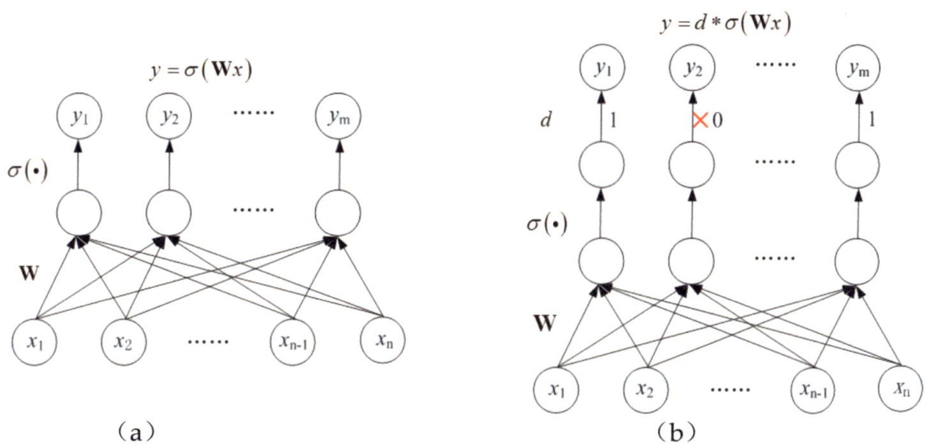

Figure 3.
Dropout method. (a) No dropout and (b) dropout.

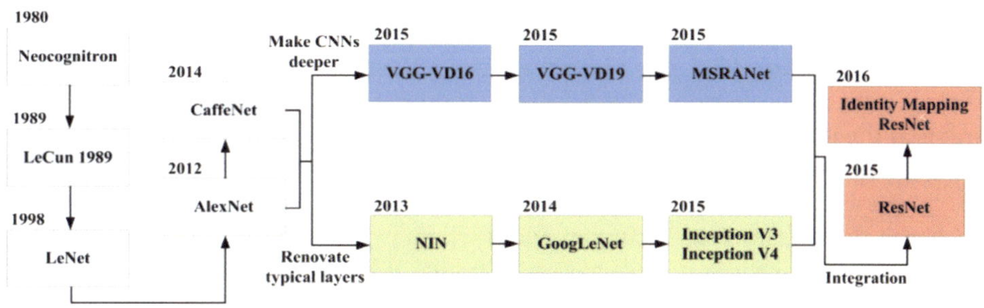

Figure 4.
Evolution of the structure of deep CNNs.

improving the speed of training, and making model combination practical, even for deep networks.

Finally, after all the convolution, pooling, and fully connected layers, a classifier layer may be used to calculate the class probability of each instance.

Based on the typical architecture of deep CNN, AlexNet [11], CaffeNet [13], VGG-VD [14], MSRA-Net [22], NIN [23], GoogLeNet [20], Inception V3 [24], Inception V4 [25], and ResNet [21] all proved to be effective in detection or classification tasks and achieve state-of-the-art performance.

In summary, we demonstrate the evolution of deep CNNs' structure in **Figure 4**:

However, these successful deep CNNs discussed above do not achieve good performance as we expected, when we directly apply them for remote scene classification. An effective solution, recently explored in [15, 18, 20], is to transfer deep features trained on ImageNet dataset to remote sensing images. Deep CNNs pre-trained by ImageNet dataset can be treated as fixed feature extractors. In a feedforward way, they extract global feature representation of the remote sensing images. With the global representation, a simple classifier can implement remote scene classification. Taking a step further, fine-tuning strategy is usually used for deeper layers of transferred deep CNNs to further improve the performance of them for remote scene classification. Typically, the first few layers are frozen, because low-level features can better fit remote scenes, and deeper layers are allowed to keep learning by training them with remote sensing images. Taking AlexNet, for example, we show the fine-tuning strategy in **Figure 5**.

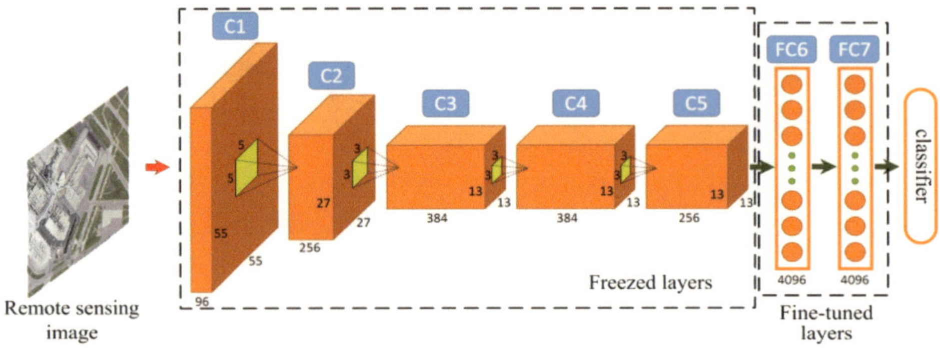

Figure 5.
Strategy of fine-tuning deeper layers of AlexNet.

Although the strategy of fine-tuning deeper layers of transferred deep CNNs with remote sensing images achieves near-perfect performance in remote scene classification [18], we challenge the theory basis of this strategy by showing that not all low-level features in shallow layers are general enough for remote scenes; some of them even shows very poor generalization power in transferring process. We find that the depth of transferred CNNs enhances the generalization power of them and guarantees a general hypothesis for remote scene classification. The detailed results are discussed in Section 3. This find in transferred deep CNNs gives an answer to the very recent discussion about whether generalization power of deep CNNs comes from sheer memorization or available hypothesis.

3. Generalization power of features in different layers of transferred deep CNN

As mentioned in Section 2, when transferring deep CNNs pre-trained by ImageNet for remote scene classification, we typically assume that features (e.g., salient edges and borders) in the shallow layers are generic, while features in the deep layers are more specific to the dataset used for pre-training and thus need to be fine-tuned by the target dataset. Therefore, the traditional strategy of transferring pre-trained deep CNNs for remote scene classification is to freeze the shallow layers and fine-tune the last deep layers. However, this assumption drives us to the question that how the "depth" of transferred deep CNNs affect the features of remote scenes in the transferring process. To answer this question, we take CaffeNet pre-trained by ImageNet, for example, and thoroughly analyze features of remote scenes in different layers of it when we transfer it for remote scene classification on UC Merced dataset[2].

Firstly, we take a close look into features of remote sensing image in the first convolutional layer of the pre-trained CaffeNet. In **Figure 6**, we visualize the convolutional filters of the first convolutional layer. These convolutional filters are learned by pre-training the CaffeNet with ImageNet dataset. We can see that the former filters are learned for extracting edges in different directions and the later filters are learned for extracting different colors. For example, the first, fifth, and ninth filters are mainly used to extract features in the right lower oblique direction,

[2] http://vision.ucmerced.edu/datasets/landuse.html

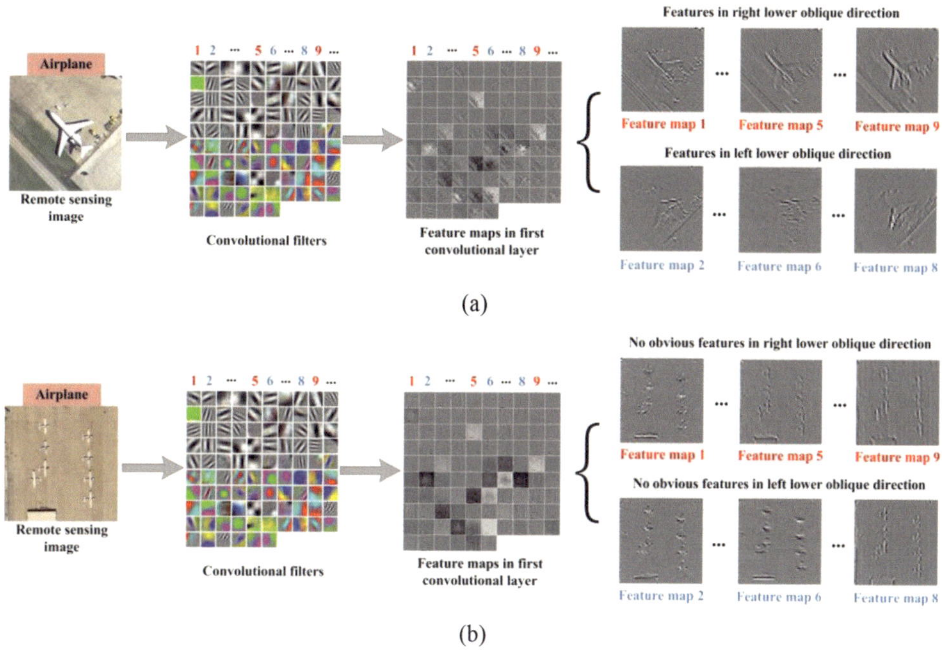

Figure 6.
Feature maps of (a) one remote sensing image and (b) another one within the same remote scene class extracted by convolutional filters in the first layer of pre-trained CaffeNet.

while the second, sixth, and eighth filters are mainly used to extract features in the left lower oblique direction. Based on the architecture of pre-trained CaffeNet, we can obtain 96 feature maps in the first convolutional layer by applying these convolutional filters to remote sensing image. In **Figure 6(a)**, we find that the first, fifth, and ninth feature maps contain features of the input image in the right lower oblique direction, while the second, sixth, and eighth feature maps contain features of the input image in the left lower oblique direction. However, in **Figure 6(b)**, we cannot see obvious features in these two directions in the corresponding feature maps. The input images in **Figure 6(a)** and **(b)** belong to the same remote scene class. However, features of them extracted by filters in the first convolutional layer of pre-trained CaffeNet are very different from each other. Compared with daily optical images in ImageNet dataset, remote sensing images are much more sophisticated. Some convolutional filters in shallow layers of pre-trained CaffeNet may be effective for some remote sensing image while affecting little about some other remote sensing images. Not all features in shallow layers of pre-trained CaffeNet are general for remote sensing images.

Furthermore, we try to visualize features of the input remote sensing image learned in deeper layers of the pre-trained CaffeNet. However, as we can see in **Figure 7**, feature maps of the remote sensing image become increasingly fuzzy from the second convolutional layer. With the increase of depth, representations of remote scene become more and more abstract. In order to reveal how the depth of pre-trained CaffeNet affects the generalization power of features in it, we intuitively reflect the distribution of features learned from the two input remote sensing images in **Figure 8** by using the t-SNE algorithm. [26, 27] In **Figure 8**, we use the t-SNE algorithm to visualize feature maps in different convolutional layers by giving each datapoint a location in a 2-D map.

Figure 8 shows the separability of features learned in different convolutional layers of pre-trained CaffeNet when we apply it on two different remote sensing

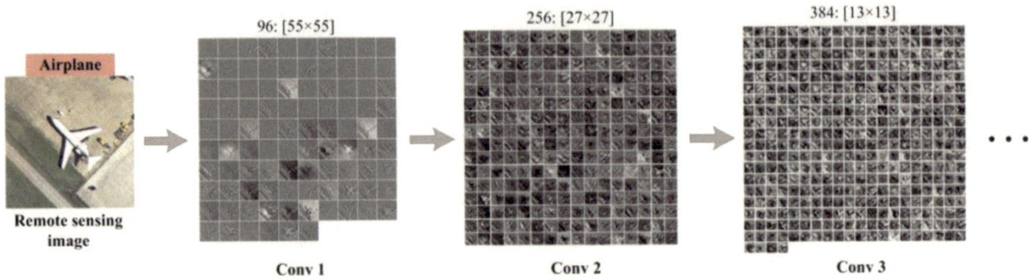

Figure 7.
Feature maps of the remote sensing image in different layers of pre-trained CaffeNet.

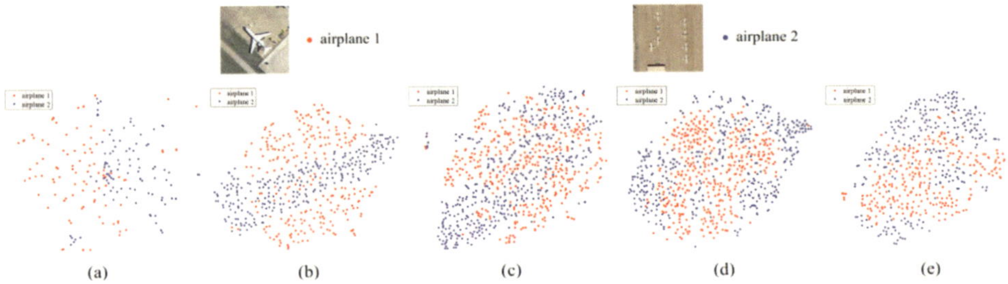

Figure 8
2-D visualization of feature maps in (a) the first convolutional layer, (b) the second convolutional layer, (c) the third convolutional layer, (d) the fourth convolutional layer, and (e) the fifth convolutional layer of pre-trained CaffeNet. The t-SNE algorithm proposed in [26, 27] is used to visualize the high-dimensional representations.

images that belong to the same remote scene class. In **Figure 8(a)**, the 2-D features of the two input images are separated to each other obviously in the first convolutional layer. Notably, from **Figure 8(a)–(e)**, the deeper the layer, the more overlap between features of the two remote sensing images we can observe. Therefore, in contrast to common belief that features in shallow layer are more generic, they are susceptible to changes in input remote sensing images. Indeed, filters of the first convolutional layer are similar to HOG, SURF, or SIFT (edge detectors, color detectors, texture, etc.). They give representative information for different input images. However, this information also conveys the specific characteristics of the dataset used to pre-train the CaffeNet. As a result, features extracted in shallow layers of pre-trained CaffeNet may be not general enough for remote scene classification in the transferring process. On the other hand, it seems that the depth of pre-trained CaffeNet enhances the generalization power of features in it. Regardless of the specific meaning of edges or colors, high-level features in deeper layer represent the sematic meaning of the input remote sensing image. Based on this analysis of features in pre-trained CaffeNet, we believe that depth of pre-trained CNNs brings general hypothesis for remote scene classification. It plays an important role when we apply pre-trained CNNs to the task of remote scene classification.

4. Experiments

The main objective of this chapter is to evaluate different deep CNNs transferred for remote scene classification. Therefore, we organize the experiments for transferred deep CNNs with various deep CNN architectures and various remote sensing datasets. We try to explore the answer for the problem where the generalization power comes from in deep CNNs and find the proper way to apply deep

CNNs for remote scene classification. All the developed codes rely on the MatConvNet[3] framework which provides a complete deep learning toolkit for training and testing models. In addition, it should be noted that all the experiments are performed on HP z820 with two Intel (R) Xeon (R) CPUs with 2.60 GHz of clock and 32GB of RAM memory. NVIDIA Quadro K2000 series is used as graphic processing units.

4.1 Experimental setup

In this section, we carry out a number of experiments based on different architectures of deep CNNs. To evaluate the effectiveness of pre-trained deep CNNs transferred for remote scene classification, we conduct experiments on three remote sensing datasets. These three datasets are different in spatial and spectral information. We compare the performance of pre-trained deep CNNs with the state-of-the-art results in these three datasets. We must note that except learning the classifier, all the experiments are unsupervised.

The three publicly available datasets used in our experiments are as follows:

UC merced land use dataset. This dataset is composed of 2100 overhead scene images divided into 21 land use scene classes. Each class consists of 100 aerial images measuring 256 × 256 pixels, with a spatial resolution of 0.3 m per pixel in the red-green-blue color space. The example images for each class are shown in **Figure 9**. This dataset was extracted from aerial orthoimagery downloaded from the United States Geological Survey (USGS) National Map of the following US regions: Birmingham, Boston, Buffalo, Columbus, Dallas, Harrisburg, Houston, Jacksonville, Las Vegas, Los Angeles, Miami, Napa, New York, Reno, San Diego, Santa Barbara, Seattle, Tampa, Tucson, and Ventura. So far, this dataset is the most popular and has been widely used for the task of remote scene classification and retrieval. [28]

WHU-RS dataset[4]. Collected from Google Earth, this dataset is composed of 950 aerial scene images with 600 × 600 pixels, which are uniformly distributed in 19

Figure 9.
One example image for each class of the UC Merced land use dataset. (a) Agricultural; (b) airplane; (c) baseball diamond; (d) beach; (e) buildings; (f) chaparral; (g) dense residential; (h) forest; (i) freeway; (j) golf course; (k) harbor; (l) intersection; (m) medium residential; (n) mobile home park; (o) overpass; (p) parking lot; (q) river; (r) runway; (s) sparse residential; (t) storage tanks; (u) tennis court.

---

[3] http://www.vlfeat.org/matconvnet/

[4] http://www.tsi.enst.fr/~xia/satellite_image_project.html.

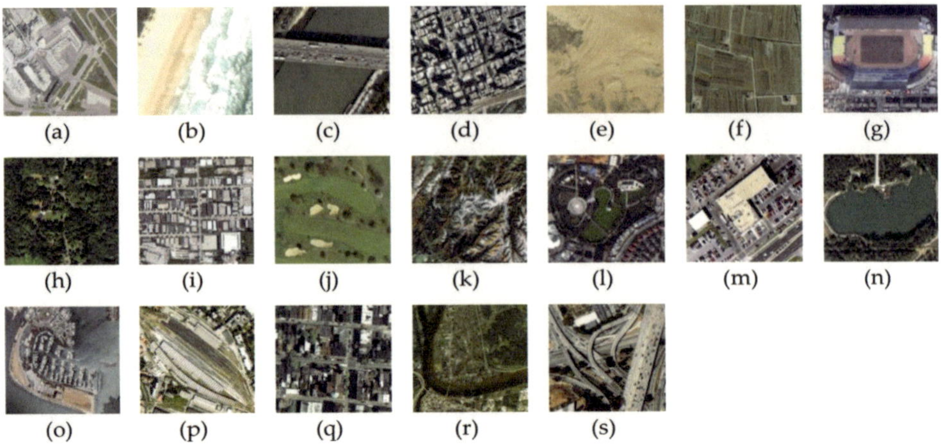

Figure 10.
One example image for each class of the WHU-RS dataset. (a) Airport; (b) beach; (c) bridge; (d) commercial; (e) desert; (f) farmland; (g) football field; (h) forest; (i) industrial; (j) meadow; (k) mountain; (l) park; (m) parking lot; (n) pond; (o) port; (p) railway; (q) residential; (r) river; (s) viaduct.

scene classes, 50 for each class. With spatial resolution up to 0.5 m and spectral bands of red, green, and blue, the example images for each class are shown in **Figure 10**. This dataset is challenging due to the high variations in resolution, scale, orientation, and illuminations of the images.

Brazilian coffee scenes dataset[5]. This dataset consists of only two scene classes (coffee class and non-coffee class), and each class has 1438 image tiles with a size of 64 × 64 pixels cropped from SPOT satellite images over four counties in the State of Minas Gerais, Brazil: Arceburgo, Guaranesia, Guaxupe, and Monte Santo. This dataset considered the green, red, and near-infrared bands because they are the most useful and representative ones for distinguishing vegetation areas. **Figure 11** shows three example images for each of the coffee and non-coffee classes in false colors.

In the experiments, we divide all the datasets in fivefolds. For UC Merced dataset, WHU-RS dataset, and Brazilian coffee scenes dataset, each of the five folds contains 420 images, 190 images, and 600 images, respectively. Then, the classification accuracy and standard deviation are calculated with fivefold cross-validation. Five well-known pre-trained deep CNNs (AlexNet [11], CaffeNet [13], VGG-VD16 [14], GoogLeNet [20], and ResNet [21]) descripted in Section 2 are used to test the effectiveness of pre-trained deep CNNs in the experiments. As we analyzed before, all the experiments are in unsupervised framework except learning the classifier.

4.2 Experiment results of remote scene classification

We evaluate transferred deep CNNs for the task of remote scene classification based on the five well-known deep CNN architectures (AlexNet, CaffeNet, VGG-VD16, GoogLeNet, and ResNet) pre-trained by ImageNet. For the strategy of transferring deep CNNs for remote scene classification, we use the five pre-trained deep CNNs to extract high-level features from input images. These input images are resized to 227 × 227 for pre-trained AlexNet and CaffeNet and 224 × 224 for pre-trained VGG-VD16, GoogLeNet, and ResNet by down-sampling or up-sampling operation. Linear SVM is used as classifier.

5 www.patreo.dcc.ufmg.br/downloads/brazilian-coffee-dataset/

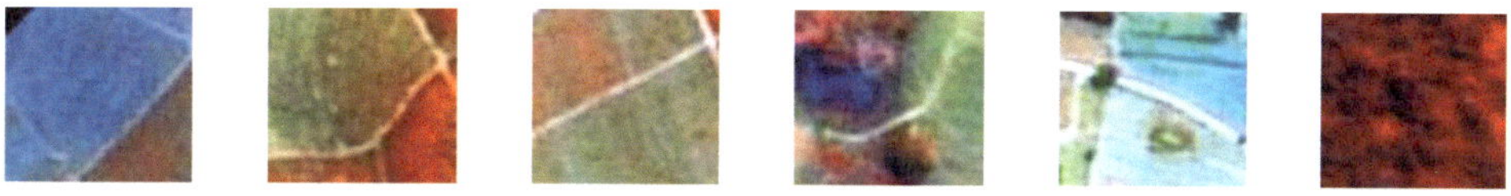

(a) (b) (c) (d) (e) (f)

Figure 11.
Example images of the Brazilian coffee scene dataset in false colors. (a)–(c) coffee class; (d)–(f) non-coffee class.

Pre-trained deep CNN	UC merced		WHU-RS		Brazilian coffee scenes	
	Ac (%)	SD	Ac (%)	SD	Ac (%)	SD
AlexNet	94.51	0.94	94.57	0.61	85.14	1.26
CaffeNet	94.12	1.05	94.67	0.75	84.97	1.54
VGG-VD16	94.43	0.68	94.76	0.72	84.12	0.97
GoogLeNet	94.57	0.98	94.68	1.01	84.06	1.16
ResNet-50	74.14	5.89	75.12	5.36	60.54	7.22
ResNet-101	72.36	5.96	72.85	5.09	59.39	6.68
ResNet-152	72.48	4.35	72.81	4.42	59.62	6.81

Table 1
Remote scene classification results of the five well-known pre-trained deep CNNs on three different remote sensing datasets.

With various pre-trained deep CNN models and remote sensing datasets, the remote scene classification performances are shown in **Table 1**. In **Table 1**, Ac and SD denote accuracy and standard deviation, respectively.

In the experiment, pre-trained deep CNNs are directly used as feature extractors in an unsupervised manner. By removing the last fully connected layer, the rest parts of pre-trained deep CNNs extract high-dimensional feature vectors of remote sensing images. These feature vectors are considered as final image representation followed by a linear SVM classifier. From **Table 1**, we can see that all transferred deep CNNs generated from AlexNet, CaffeNet, VGG-VD16, and GoogLeNet achieve state-of-the-art performance. Pre-trained deep CNNs show strong generalization power in the transferring process. In addition to our surprise, the most successful deep CNNs to date, ResNets fail to obtain a good experiment result, no matter their layers are 50, 101, or 152. In ResNets, shortcut connections bring less parameters and make the network much easier to optimize. At the same time, the direct connection between input and output brings poor generalization ability when we transfer them for other tasks. On the other hand, as shown in **Figure 11**, the spatial information of remote sensing images in the Brazilian coffee scene dataset is very simple. However, these remote sensing images are not optical (green-red-infrared). In **Table 1**, the relatively poor performance on this dataset comes from the difference in spectral information when we are transferring pre-trained deep CNNs for remote scene classification.

In order to test the performance of transferred deep CNNs for each remote scene class, in **Figure 12**, we draw the confusion matric of the experiment results on UC Merced dataset based on pre-trained CaffeNet.

In **Figure 12**, the experiment results in perfect or near-perfect accuracy for most of the scene categories. The relatively lower classification accuracy lies in the categories of building, dense residential, medium residential, and tennis court.

However, all these classes have some very "close" neighbors. Taking dense residential as example, it suffers the presence of very close classes, like buildings and medium residential, which we cannot even distinguish by eyes. Taking pre-trained CaffeNet, for example, **Figure 13** shows the detail changes of an optical remote sensing image.

Abbreviated as "conv" and "fc," reconstructions of convolutional feature maps in the former network layers and that of fully connected layers are shown in **Figure 13**. **Figure 13** shows that the representations of convolutional layers are still photographically similar with the remote sensing image to some extent, although they become fuzzier and fuzzier from "conv1" to "conv5." In addition, the fully connected layers rearrange the information from lower layers to generate representations that are more abstract. They compose of parts (e.g., the wings of airplanes) similar but not identical to the ones found in the original image.

In **Table 2**, we compare our best result achieved via transferred deep CNNs with various state-of-the-art methods on the UC Merced dataset. With a straightforward and simple framework, transferred deep CNN achieves outstanding performance on this dataset. We must note that our proposed method just provides basic

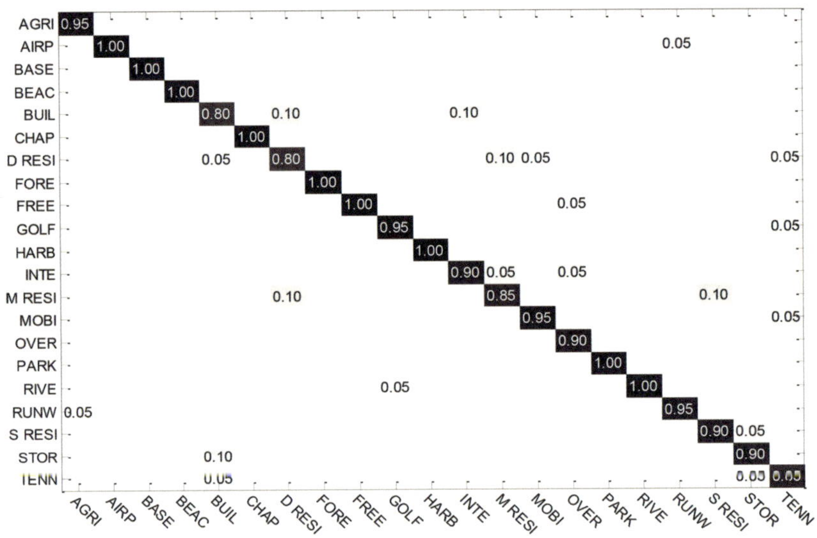

Figure 12.
Confusion matrices of classification accuracies on UC Merced dataset based on pre-trained CaffeNet.

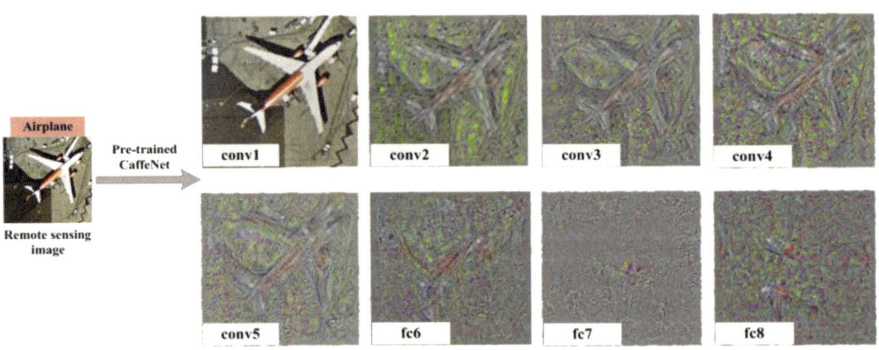

Figure 13.
Reconstruction of deep CNN activations from different layers of transferred CaffeNet. The method presented in [29] is used for visualization.

Method	Reference	Accuracy (%)
SCK	[28]	72.52
SPCK++	[30]	77.38
BRSP	[31]	77.80
UFL	[5]	81.67
CCM-BOVW	[32]	86.64
mCENTRIST	[33]	89.90
MSIFT	[34]	90.97
COPD	[35]	91.33
Dirichlet	[36]	92.80
VLAT	[10]	94.30
MCMI-based	[37]	88.20
PSR	[38]	89.10
UFL-SC	[39]	90.26
Partlets	[40]	91.33
Sparselets	[41]	91.46
FBC	[42]	85.53
LPCNN	[43]	89.90
MTJSLRC	[44]	91.07
SSBFC	[45]	91.67
CTS	[46]	93.08
Transferred GoogLeNet	—	**94.57**

Table 2
Classification accuracy (%) of reference and transferred deep CNN on the UC Merced dataset.

framework to directly transfer pre-trained deep CNNs for remote scene classification in an unsupervised manner. The effectiveness of fine-tuning approach is much dependent on the amount of images in remote sensing dataset, and the computation time of it is more demanding compared with our proposed strategy [15].

5. Discussion

To solve the problem that deep CNNs tend to over-fit when trained with limited remote sensing dataset, generalization power of deep CNNs plays the key role. In this chapter, we try to transfer deep CNNs pre-trained by daily images for remote scene classification and provide an insight for the generalization power of features in the transferred deep CNNs. From the extensive experiments above, the deep architecture of CNNs, which extracts semantic features of remote scenes, has been proven to be critical for remote scene classification. Specifically, several practical observations from the experiments and some limitations of our study are summarized as follows:

From **Table 1**, we can see that with our proposed method the classification accuracies of UC Merced dataset and WHU-RS dataset can both achieve state-of-the-art results which are near 95%. In addition, small standard deviation of classification accuracy suggests that our proposed method is stable when applied for remote scene classification. To our surprise, the most successful deep CNNs to date, ResNets, fail to obtain good experiment result when we transfer it for remote scene classification, no matter their layers are 50, 101, or 152. Shortcut connections in ResNets bring poor generalization ability when we transfer them to remote scenes. [21] This phenomenon indicates that not all successful deep CNNs are suitable for transferring to the task of remote scene classification.

Different from the traditional view that all basic features (e.g., salient edges and borders) in shallow layers of a deep CNN are more general than that learned in deep layers, we find some features in shallow layer of deep CNNs show poor generalization power when we transfer them for remote scene classification. High-level features learned in deeper layers of transferred deep CNNs are more general than these basic features.

In the remote sensing field, the scale of remote sensing datasets will be larger and larger. On the other hand, the structure of deep CNN will be optimized, and the parameters in it will be less and less. [47] Therefore, we could get more and more useful information from remote sensing datasets, which provide a priori knowledge for pre-trained deep CNNs and result in better generalization power.

Based on our study, the future research directions of applying deep CNNs for remote scene classification may be as follows. Firstly, as we discussed above, when transferring the most successful ResNet for remote scene classification, it does not work as we expected. What is the proper architecture of deep CNN that is suitable to transfer to remote scenes? Secondly, instead of directly transferring pre-trained deep CNNs for remote scene classification, could we replace some basic features that show poor generalization power in shallow layers of transferred deep CNN? Finally, with more and more remote sensing information coming into our sight, how can we use these a priori knowledge when we apply deep CNNs for remote scene classification?

6. Conclusion

In this chapter, we have presented a framework to investigate the effectiveness of transferred deep CNNs for remote scene classification. We test transferred deep CNNs for different remote sensing datasets and take a close look into the generalization power of features in them.

The two main conclusions of this work are that (1) without shortcut connections in the deep architecture as ResNet dose, most CNNs transferred from well-known pre-trained deep CNNs achieve state-of-the-art performance in remote scene classification. (2) We further confirm the conclusion in the background of remote scene classification that the generalization power derived from deep architectures brings general hypothesis. Compared with basic features (e.g., salient edges and borders), features in deeper layers are more general for remote scenes. Experiments on three remote sensing datasets with different image resolutions have provided insightful information. Transferred deep CNN improves the classification accuracy of remote scenes on UC Merced dataset with a gain up to 1.49% compared with other methods. High-level feathers in deeper layers of transferred deep CNNs are more general for remote scene classification and result in satisfied performance in unsupervised setting.

We believe our work in this chapter provides a thorough analysis about the generalization power of transferred deep CNNs for remote scene classification. It can serve as a good baseline for people to apply deep CNNs to other remote sensing datasets.

Acknowledgements

This work was supported by the National Natural Science Foundation of China under Grant No. 61601499 and No. 61601505. All the funds above can cover the costs to publish in open access.

Conflict of interest

The authors declare no conflict of interest. The founding sponsors had no role in the design of the study; in the collection, analyses, or interpretation of data; in the writing of the manuscript, and in the decision to publish the results.

Author details

Chang Luo[1*†], Hanqiao Huang[2†], Yong Wang[1] and Shiqiang Wang[3]

1 Troops of 78092, Cheng Du, China

2 Unmanned System Research Institute, Northwestern Polytechnical University, Xi'an, China

3 Air and Missile Defense College, Air Force Engineering University, Xi'an, China

*Address all correspondence to: luochang1988@126.com

† These authors contributed equally to this work and should be considered co-first authors.

References

[1] Wang J, Qin Q, Li Z, et al. Deep hierarchical representation and segmentation of high resolution remote sensing images. In: 2015 IEEE International Geoscience and Remote Sensing Symposium (IGARSS). IEEE; 2015. pp. 4320-4323

[2] Nijim M, Chennuboyina RD, Al AW. A supervised learning data mining approach for object recognition and classification in high resolution satellite data. International Journal of Computer, Electrical, Automation, Control and Information Engineering. 2015;**9**(12):2319-2323

[3] Vakalopoulou M, Karantzalos K, Komodakis N, et al. Building detection in very high resolution multispectral data with deep learning features. In: 2015 IEEE International Geoscience and Remote Sensing Symposium (IGARSS). IEEE; 2015. pp. 1873-1876

[4] Zhou W, Shao Z, Diao C, et al. High-resolution remote-sensing imagery retrieval using sparse features by auto-encoder. Remote Sensing Letters. 2015; **6**(10):775-783

[5] Cheriyadat AM. Unsupervised feature learning for aerial scene classification. IEEE Transactions on Geoscience and Remote Sensing. 2014; **52**(1):439-451

[6] Xu Y, Huang B. Spatial and temporal classification of synthetic satellite imagery: Land cover mapping and accuracy validation. Geo-spatial Information Science. 2014;**17**(1):1-7

[7] Yang W, Yin X, Xia GS. Learning high-level features for satellite image classification with limited labeled samples. IEEE Transactions on Geoscience and Remote Sensing. 2015; **53**(8):4472-4482

[8] Shao W, Yang W, Xia GS. Extreme value theory-based calibration for the fusion of multiple features in high-resolution satellite scene classification. International Journal of Remote Sensing. 2013;**34**(23):8588-8602

[9] Romero A, Gatta C, Camps-Valls G. Unsupervised deep feature extraction for remote sensing image classification. IEEE Transactions on Geoscience and Remote Sensing. 2016;**54**(3):1349-1362

[10] Negrel R, Picard D, Gosselin PH. Evaluation of second-order visual features for land-use classification. In: 2014 12th International Workshop on Content-Based Multimedia Indexing (CBMI). IEEE; 2014. pp. 1-5

[11] Krizhevsky A, Sutskever I, Hinton GE. Imagenet classification with deep convolutional neural networks. In: Advances in Neural Information Processing Systems. 2012. pp. 1097-1105

[12] Sermanet P, Eigen D, Zhang X, et al. Overfeat: Integrated recognition, localization and detection using convolutional networks. arXiv preprint arXiv:1312.6229; 2013

[13] Jia Y, Shelhamer E, Donahue J, et al. Caffe: Convolutional architecture for fast feature embedding. In: Proceedings of the 22nd ACM International Conference on Multimedia. ACM; 2014. pp. 675-678

[14] Simonyan K, Zisserman A. Very deep convolutional networks for large-scale image recognition. arXiv preprint arXiv:1409.1556; 2014

[15] Castelluccio M, Poggi G, Sansone C, et al. Land use classification in remote sensing images by convolutional neural networks. arXiv preprint arXiv: 1508.00092; 2015

[16] Deng J, Dong W, Socher R, et al. Imagenet: A large-scale hierarchical

image database. In: Computer Vision and Pattern Recognition, 2009. CVPR 2009. IEEE; 2009. pp. 248-255

[17] Nanni L, Ghidoni S. How could a subcellular image, or a painting by van Gogh, be similar to a great white shark or to a pizza? Pattern Recognition Letters. 2017;**85**:1-7

[18] Penatti OAB, Nogueira K, dos Santos JA. Do deep features generalize from everyday objects to remote sensing and aerial scenes domains? In: Proceedings of the IEEE Conference on Computer Vision and Pattern Recognition Workshops. 2015. pp. 44-51

[19] Hu F, Xia G-S, Hu J, et al. Transferring deep convolutional neural networks for the scene classification of high-resolution remote sensing imagery. Remote Sensing. 2015;7:14680-14707

[20] Szegedy C, Liu W, Jia Y, et al. Going deeper with convolutions. In: Proceedings of the IEEE Conference on Computer Vision and Pattern Recognition. 2015. pp. 1-9

[21] He K, Zhang X, Ren S, et al. Deep residual learning for image recognition. arXiv preprint arXiv:1512.03385; 2015

[22] He K, Sun J. Convolutional neural networks at constrained time cost. In: Proceedings of the IEEE Conference on Computer Vision and Pattern Recognition. 2015. pp. 5353-5360

[23] Lin M, Chen Q, Yan S. Network in network. arXiv preprint arXiv: 1312.4400; 2013

[24] Szegedy C, Vanhoucke V, Ioffe S, et al. Rethinking the inception architecture for computer vision. arXiv preprint arXiv:1512.00567; 2015

[25] Szegedy C, Ioffe S, Vanhoucke V. Inception-v4, inception-resnet and the impact of residual connections on learning. arXiv preprint arXiv: 1602.07261; 2016

[26] Maaten L, Hinton G. Visualizing data using t-SNE. Journal of Machine Learning Research. 2008;**9**(Nov): 2579-2605

[27] Van Der Maaten L. Accelerating t-SNE using tree-based algorithms. Journal of Machine Learning Research. 2014;**15**(1):3221-3245

[28] Yang Y, Newsam S. Bag-of-visual-words and spatial extensions for land-use classification. In: Proceedings of the 18th SIGSPATIAL International Conference on Advances in Geographic Information Systems. ACM; 2010. pp. 270-279

[29] Mahendran A, Vedaldi A. Understanding deep image representations by inverting them. In: 2015 IEEE Conference on Computer Vision and Pattern Recognition (CVPR). IEEE; 2015. pp. 5188-5196

[30] Yang Y, Newsam S. Spatial pyramid co-occurrence for image classification. In: 2011 International Conference on Computer Vision. IEEE; 2011. pp. 1465-1472

[31] Jiang Y, Yuan J, Yu G. Randomized spatial partition for scene recognition. In: Computer Vision–ECCV 2012. Berlin, Heidelberg: Springer; 2012. pp. 730-743

[32] Zhao LJ, Tang P, Huo LZ. Land-use scene classification using a concentric circle-structured multiscale bag-of-visual-words model. Journal of Selected Topics in Applied Earth Observations and Remote Sensing. 2014;7(12):4620-4631

[33] Xiao Y, Wu J, Yuan J. mCENTRIST: A multi-channel feature generation mechanism for scene categorization. IEEE Transactions on Image Processing. 2014;**23**(2):823-836

[34] Avramović A, Risojević V. Block-based semantic classification of high-resolution multispectral aerial images. Signal, Image and Video Processing. 2016;**10**(1):75-84

[35] Cheng G, Han J, Zhou P, et al. Multi-class geospatial object detection and geographic image classification based on collection of part detectors. ISPRS Journal of Photogrammetry and Remote Sensing. 2014;**98**:119-132

[36] Kobayashi T. Dirichlet-based histogram feature transform for image classification. In: Proceedings of the IEEE Conference on Computer Vision and Pattern Recognition. 2014. pp. 3278-3285

[37] Ren J, Jiang X, Yuan J. Learning LBP structure by maximizing the conditional mutual information. Pattern Recognition. 2015;**48**(10):3180-3190

[38] Chen S, Tian YL. Pyramid of spatial relations for scene-level land use classification. IEEE Transactions on Geoscience and Remote Sensing. 2015; **53**(4):1947-1957

[39] Hu F, Xia GS, Wang Z, et al. Unsupervised feature learning via spectral clustering of multidimensional patches for remotely sensed scene classification. Journal of Selected Topics in Applied Earth Observations and Remote Sensing. IEEE; 2015;**8**(5):13

[40] Cheng G, Han J, Guo L, et al. Effective and efficient midlevel visual elements-oriented land-use classification using VHR remote sensing images. IEEE Transactions on Geoscience and Remote Sensing. 2015; **53**(8):4238-4249

[41] Cheng G, Han J, Guo L, et al. Learning coarse-to-fine sparselets for efficient object detection and scene classification. In: Proceedings of the IEEE Conference on Computer Vision and Pattern Recognition. 2015. pp. 1173-1181

[42] Hu F, Xia GS, Hu J, et al. Fast binary coding for the scene classification of high-resolution remote sensing imagery. Remote Sensing. 2016;**8**(7):555

[43] Zhong Y, Fei F, Zhang L. Large patch convolutional neural networks for the scene classification of high spatial resolution imagery. Journal of Applied Remote Sensing. 2016;**10**(2): 025006-025006

[44] Kunlun Qi, Wenxuan Liu, Chao Yang, et al. High resolution satellite image classification using multi-task joint sparse and low-rank representation. Preprints, 7 November 2016, doi:10.20944/preprints201611.0036.v1, (www.preprints.org)

[45] Zhao B, Zhong Y, Zhang L. A spectral–structural bag-of-features scene classifier for very high spatial resolution remote sensing imagery. ISPRS Journal of Photogrammetry and Remote Sensing. 2016;**116**:73-85

[46] Yu H, Yang W, Xia GS, et al. A color-texture-structure descriptor for high-resolution satellite image classification. Remote Sensing. 2016; **8**(3):259

[47] LeCun Y, Bengio Y, Hinton G. Deep learning. Nature. 2015;**521**:436-444

Chapter 5

Sub-Pixel Technique for Time Series Analysis of Shoreline Changes Based on Multispectral Satellite Imagery

Qingxiang Liu and John C. Trinder

Abstract

The measurement and monitoring of shoreline changes are of great interest to coastal managers and engineers. Shoreline change information can be crucial for the assessment of coastal disasters, design of coastal infrastructure and protection of coastal environment. This chapter presents shoreline change monitoring based on multispectral satellite imagery and sub-pixel technique. Firstly, a brief introduction of shoreline definitions and indicators is given. Sub-pixel techniques for shoreline mapping on multispectral satellite images are then introduced. Following that, a brief review of existing research studies of long-term shoreline change monitoring based on multispectral imagery is given. Subsequently, a case study of sub-pixel shoreline change monitoring at the northern Gold Coast on the east coast of Australia is presented. By comparing the longshore averaged beach widths at seven representative transects from Landsat with those from Argus imaging data, the RMSEs range from 9.1 to 12.3 m and the correlations are all no less than 0.7. Annual means and variabilities of beach widths were estimated without significant differences from the reference data for most of the results. Finally, conclusions and recommendations for future work are given.

Keywords: shoreline mapping, change monitoring, satellite imagery, multispectral, super-resolution mapping (SRM)

1. Introduction

1.1 Shoreline definitions and indicators

A 'Shoreline' is ideally defined as the interface between the land and water [1]. However, because of its temporally and spatially dynamic nature, the definition of shoreline should be considered in a temporal sense [2]. The shoreline is dynamic as unconsolidated sediments along the beach adjust constantly to changes of environmental forces. Sediments are deposited offshore during energetic conditions, resulting in an erosion trend of the shoreline; during calm conditions, sediments are returned back to the subaerial area, leading to an accretion of the shoreline [3]. The erosion-recovery circles may occur over several days, or as long as several decades during extreme wave conditions. In addition to shoreline changes resulting from cross-shore sediment exchanges, along sediment exchanges can also occur, leading to relative movements between updrift and downdrift ends of the beach, which is commonly referred to as "beach rotation" [3].

For practical purposes, coastal investigators have adopted various indicators to approximate the real shoreline positions. A comprehensive literature review of shoreline definitions with extensive references is given in [2]. Three groups of shoreline indicators exist. The first group of shoreline indicators are based on visually interpreted features. A visually interpreted shoreline is a coastal feature that can be physically seen, which can change without any onshore or offshore movement of sand because it is affected by tide, wave and weather conditions [4]. Despite the definitions, some visually interpreted shoreline indicators may sometimes not be observable. For example, the high water line (HWL), which was defined as the mark left by maximum run-up from the previous high tide, may appear as a transitional zone instead of a clear line, or may not be visible at all [5]. Also, as the sand gradually dries out after previous waves, a discernible wet/dry boundary may not be available. In addition, some of the indicators are ambiguous, subjective or even inconsistent between different studies [2]. Therefore, visually interpreted shorelines may require experience and skills from the interpreters.

The second group of shoreline indicators are tidal datum-based, which are determined by the intersection of the beach profile with specific vertical tidal elevations. Generally, they are temporally and spatially more stable and are less responsive to wave condition changes than visually interpreted shorelines [4, 6, 7]. Tidal datum-based shorelines are easy to understand and less likely to be ambiguous, but their extraction requires beach profile data, i.e. 3D survey data.

The third group of shoreline indicators reported by [2] depend purely on image processing techniques and are not necessarily visually discernible. As digital image processing is more and more widely used, especially when an automatically instead of manually extracted shoreline is required, this group of shoreline indicators are not uncommon nowadays, e.g. [8–10].

The decision as to which group of shoreline indicators to use largely depends on data availability. Taking 2D satellite imagery as an example, the lack of elevation information prohibits the extraction of tidal datum-based shorelines. Also, the spatial resolution of the images significantly affects the feasibility of extracting visually discernible shorelines. Since the stabilities of these shoreline indicators are different, the decision should also take account of the investigated temporal scales, i.e. whether short-term or long-term shoreline change is of interest. In addition, a single shoreline indicator should be consistently used as possible when comparing shoreline changes over time.

1.2 Sub-pixel techniques for shoreline mapping

While shorelines are primarily linear features, the extraction using only edge detection methods can be a difficult task, because of the lack of sufficient and consistent contrast between water body and land. Besides, shorelines should be continuous and normally unique, which increases the requirements of extracting edges. In this case, shoreline mapping can be converted to the task of boundary extraction between water and land regions on a labelled map, which is produced through a classification step.

For images with medium to low spatial resolutions, pixels are most likely to be mixed. In other words, many pixels contain multiple land covers and the actual boundaries between classes generally run through the content of mixed pixels. Consequently, a traditional pixel-based classification which forces each pixel to be assigned to a single class label may mis-locate the boundaries between classes. This mis-location will be more significant as the pixel size increases. In this case, image processing at a sub-pixel level is preferred. Sub-pixel mapping techniques, which is also referred to as super-resolution mapping (SRM), was defined as the process of spatially designating class proportions to concrete pure sub-pixels [11]. SRM aims to produce a classification map at a finer scale than the original image, under the basic assumption of spatial dependence between pixels [12]. According to [11], there are generally two steps for SRM as shown in **Figure 1**. The first step is calculating the proportion of each class inside a pixel, which is also called soft classification. The second step is spatially assigning land cover classes to pure sub-pixels according to their proportions in the pixel, after which a classification map at a finer scale is generated. Readers are referred to [11] for more details of SRM.

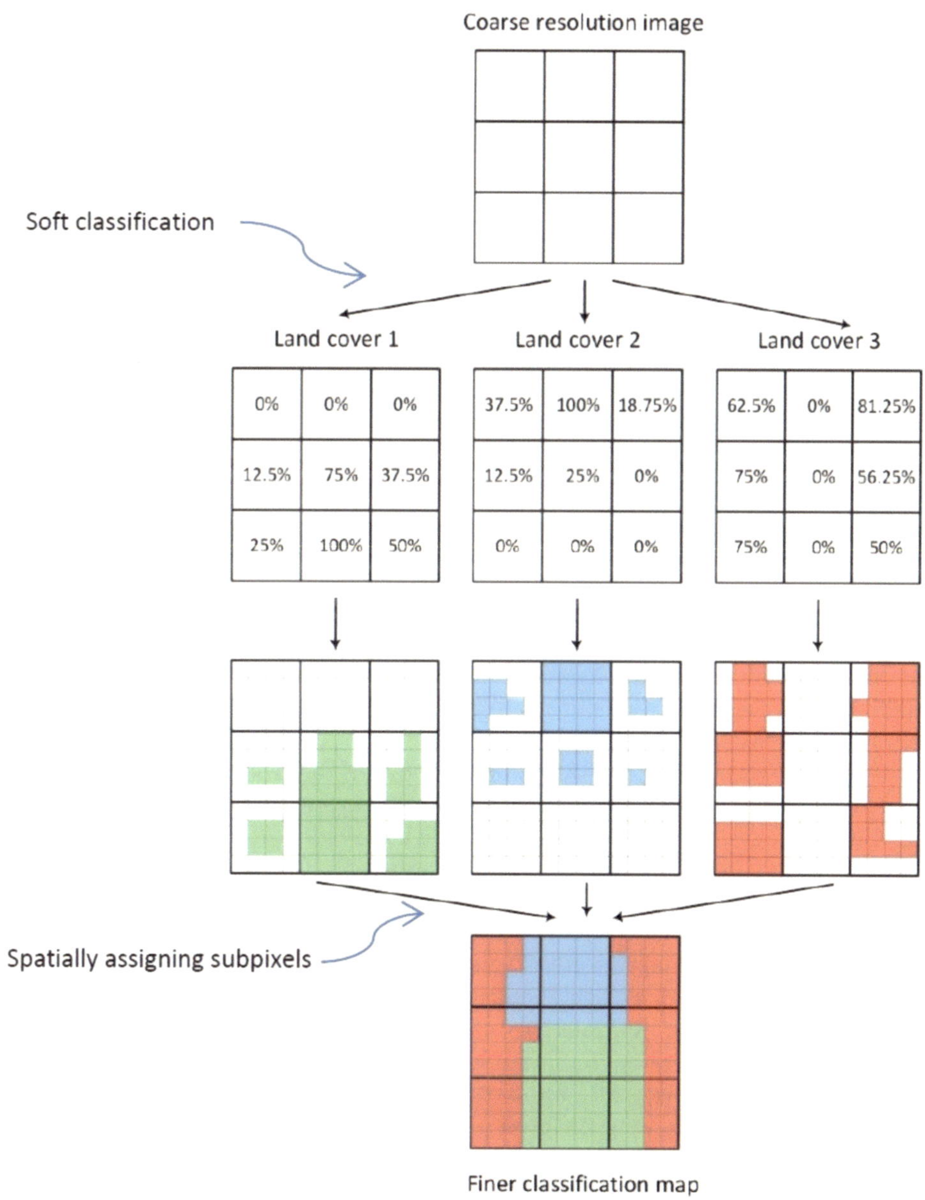

Figure 1.
SRM principle (adapted from [11, 15]).

SRM has been demonstrated in some research works (e.g. [11–14]) to be suitable for image classification at sub-pixel level.

Ref. [16] compared the performance of three SRM methods, i.e. contouring of soft classification [17], wavelet based interpolation and geo-statistical two-point histogram methods [18], for the extraction of sub-pixel shorelines from a degraded IKONOS satellite image. They used the shoreline extracted from the original IKONOS image as reference and demonstrated that all three SRM methods outperformed pixel-based classification and that the geostatistical two-point histogram method gave the best result over the study site. Ref. [19] tested two SRM methods also on a degraded IKONOS image, using both local and global training statistics. In Ref. [20] Normalised Difference Water Index (NDWI) was calculated to distinguish water and land, which was defined as

$$NDWI = \frac{B_{Green} - B_{NIR}}{B_{Green} + B_{NIR}} \qquad (1)$$

where B_{Green} and B_{NIR} represent green and NIR bands of Landsat Enhanced Thematic Mapper Plus (ETM+) images respectively. Lake shorelines were then extracted at sub-pixel level based on sub-pixel edge localisation and smoothing. Using QuickBird panchromatic images as reference data, the authors demonstrated that the extracted shorelines gave a better estimation of lake areas and perimeters than pixel-based results. Ref. [9] developed a method to automatically extract sub-pixel shorelines using the near-infrared (NIR) band of Landsat images. They compared 45 sub-pixel shorelines with manually edited shorelines from aerial photos, demonstrating that the error of sub-pixel shoreline locations over their study site is less than 6 m. All of these above mentioned studies indicate that SRM techniques can be effective for improving shoreline mapping accuracy on medium to low-resolution satellite images.

1.3 Long-term shoreline change monitoring using multispectral sa tellite images

Since most of the high-resolution satellites were launched no earlier than two decades ago, they are mostly not suitable for monitoring long-term (e.g. multi-decadal) shoreline changes. Therefore, most of the studies in the literature used medium to low-resolution satellite data such as Landsat, ASTER and SPOT imagery as the main data source, e.g. [21–23].

The Landsat program dates back to 1972, producing multi-decadal archival data freely available to the public with a revisit time of 16 days since Landsat-4. Although a number of case studies on using Landsat data to monitor long-term shoreline changes exist, there are a limited number of existing long-term coastal monitoring programs producing ground-truth data with both high spatial and temporal resolutions [24, 25]. The availability of ground-truth data is likely to have limited the amount and temporal frequency of experimental data used in the literature. Most of the studies have used only a small percentage of available archival Landsat satellite images with yearly frequency or even longer time intervals, e.g. [26–28]. Only a few studies have used a frequency higher than yearly [9, 29] or even all the available Landsat data [30, 31] over the studied areas. However, the ground truth data used in these studies was either over a much shorter term [30] or much lower than monthly frequency [31]. Consequently, intra-annual variability over a long term may not be revealed and validated. Ref. [10] presented a case study of using full-frequency archival Landsat data for monitoring of shoreline changes during approximately three decades, where long-term and temporally dense ground surveying data is used as reference data.

2. Long-term shoreline change monitoring: a case study at the northern Gold Coast, Australia

2.1 Study site description

The Gold Coast is a coastal city located approximately 66 km southeast of Brisbane near the Queensland-New South Wales state border (**Figure 2(b)** and **(c)**). The Gold Coast comprises a 35 km-long east-facing shoreline of over a dozen of beaches, stretching from the Rainbow Bay at the southern end to South Stradbroke Island in the north. The study area is a part of the northern Gold Coast between Main Beach and Broadbeach (**Figure 2(a)**), extending 4.5 km alongshore.

The Gold Coast beaches are characterised as energetic, intermediate beaches with mean offshore significant wave height and peak wave period of 1.1 m and 9.4 s respectively [32]. Beach sediments of the site have a mean grain size of 0.25 mm and a fall velocity of around 0.03 m/s. Tides in this area are micro-tidal and semidiurnal with a mean spring tidal range of approximately 1.8 m. Waves are predominantly from the southeast and show strong seasonal variations, with larger waves and more frequent storms during Australian summer to fall months (i.e. December to June) while milder waves occur during winter and spring months [32]. The nearshore morphology of the study site is characterised as a double sandbar system with a nearshore bed slope around 0.02 [33]. The net longshore drift is estimated to be 500,000 m^3 per year northwards [32]. This study site is an open, straight sandy beach, which is a representative of moderate to high wave energy, wave-dominated (micro-tidal) beaches.

2.2 Northern Gold Coast Beach protection strategy

The Gold Coast is a major tourist destination that offers some of the most popular surfing beaches in Australia. However the tourist economy is at risk of

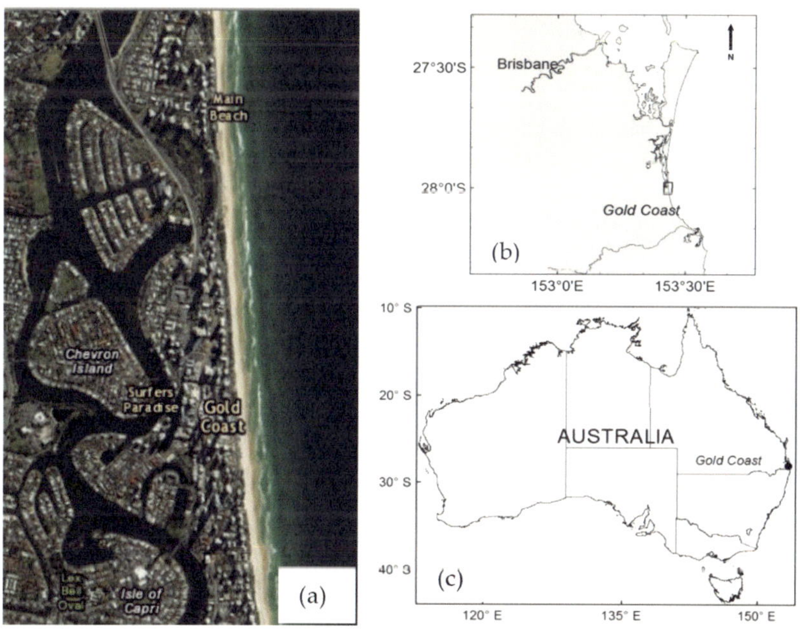

Figure 2.
(a) Study site: the northern Gold Coast (source: ArcMap basemap). (b) The location of Gold Coast with respect to Brisbane. The study area is marked as the small black rectangle. (c) The location of Gold Coast in the map of Australia.

significant losses due to beach erosion during major storm events. The Northern Gold Coast Beach Protection Strategy (NGCBPS) was a long-term, sustainable plan to maintain and enhance the beaches at the northern Gold Coast [34]. As a part of the NGCBPS, over 1.2 million cubic metres of sediment were deposited on the northern Gold Coast beaches and within the nearshore to increase beach amenity and widen the beaches as preparation for future storm events. The locations of the six sand nourishment deposition areas A1 to A6 are indicated in **Figure 3(a)**.

The beach nourishment program was commenced in February 1999 and the major phase was completed in June 2000. The progress of the nourishment is shown in **Figure 3(b)**, where the cumulative nourishment volumes at the six deposition areas are indicated using different grey scales. Note that A1a is an area approximately 300 m north of A1 and is not shown in **Figure 3(a)**. In addition, a submerged artificial reef (marked as a blue circle in **Figure 3(a)**) was built within the nearshore at Narrowneck primarily to stabilise the beach nourishment and improve surfing quality, which was initiated approximately a half year later than the commencement of the beach nourishment (in August 1999). The main phase of the reef construction was completed in December 2000.

2.3 Data used

2.3.1 Reference shorelines

Funded by the City of Gold Coast, an Argus coastal monitoring system was installed in late July 1999 at the northern Gold Coast by the Water Research Laboratory (WRL), University of New South Wales to monitor shoreline changes. Four video cameras were installed approximately 100 m above the ground on the

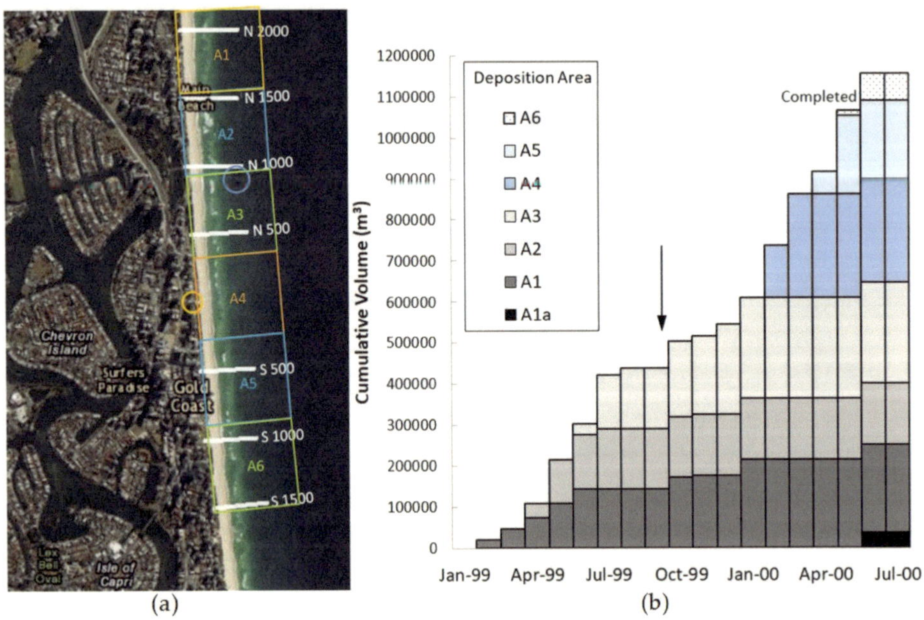

Figure 3.
(a) Locations of sand nourishment deposition areas A1 to A6 (marked as squares). The locations of the focus building where the Argus cameras were installed and the artificial reef at Narrowneck are marked as orange and blue circles respectively. The locations of the seven regularly spaced cross-shore profile lines, i.e. from 2000 m north (N 2000) to 1500 m south (S 1500) of the Argus station are also indicated. (b) Progress of sand nourishment (adapted from [35]) at all deposition areas. The black arrow indicates the time when the Argus coastal monitoring was initiated.

Focus Building, which is located about 60 m landward of the dune line and 900 m to the south of Narrowneck (**Figure 3(a)**). The system became fully operational in August 1999, which coincided with the commencement of Narrowneck reef construction. The Argus monitoring project lasted from 1999 to 2008 and was recommenced in 2014.

Every daylight hour, the cameras collected a 10-minute time-averaged image, from which the natural variations of breaking waves were effectively averaged. To minimise the effects of tidal variations, only the time-averaged images acquired at mid-tide, which is approximately 0 m on the Australian Height Datum (AHD) were selected. Among the selected images, only those acquired when the root-mean-square wave heights were no more than 1 m were retained for future use. These images were rectified and geo-referenced through a standardised image pre-processing procedure [35]. Shorelines were then extracted by WRL staff using the pixel intensity clustering (PIC) method [8, 36]. The shoreline positions were defined as the cross-shore locations of the mean sea level contours [37] and may not be visually discernible on the time-averaged video images. Therefore, they are believed to belong to the third group of shoreline indicators and depend on the PIC method. It is estimated that the horizontal errors of the extracted shoreline time-series were within ±5 m [37]. While the nominal frequency of the shorelines is weekly [35], the actual frequency depends on the wave conditions during the monitoring period. During the studied period (1999 to 2008), the average time frequency of available shorelines is 6–9 days.

2.3.2 Experimental Landsat images

Decided by the availability of the reference data, Landsat multispectral images spanning approximately 9 years, i.e. from August 1999 to October 2008 were used. The dataset consists of images acquired by two satellite instruments—Landsat 5 Thematic Mapper (TM) and Landsat 7 ETM+. For Landsat 7 ETM+ instrument, the products before and after the scan line corrector (SLC) failure are used. Surface Reflectance products which have been atmospherically corrected based on L1T (precision terrain) data were ordered from U.S. Geological Survey (USGS) website https://espa.cr.usgs.gov/. Radiometric calibration and cross-calibration have been implemented [38] by USGS. As reported by NASA, the geo-registration errors are within 0.4 pixels (12 m), indicating the variation in position between images at different times is at subpixel level. The numbers and spanning periods of images used, after eliminating scenes blocked by thick clouds, are indicated as **Table 1**.

2.4 Shoreline extraction from Landsat images

The shoreline extraction and change monitoring flow chart is described as **Figure 4**. Firstly, Landsat images were clipped based on the boundary of the study site. A simple one-dimensional cubic interpolation was employed to fill the gaps on the Landsat 7 images in SLC-off mode. This step was merely for extracting continuous shorelines and the interpolated values were not utilised for beach widths calculation. Then, shorelines at sub-pixel level were extracted from the nine-year Landsat data using the strategy illustrated in [10]. Due to the limitation of the spatial resolution of Landsat images, the extracted shorelines are not visually discernible. Therefore, theses shorelines also belong to the third group of shoreline indicators and depend on the shoreline extraction strategy.

Subsequently, tidal correction was applied to the shorelines extracted from Landsat images. Assuming the beach face slope did not vary significantly alongshore, which is a valid judgement from the averaged historical ground measurements, the horizontal and cyclic translation of the shoreline caused by tidal variation can be simplified as

$$\Delta = Z/m, \quad (2)$$

where Δ is the horizontal cross-shore shift; z is the tidal elevation at image acquisition time relative to 0 m AHD. Astronomical tidal elevations recorded every 15 minutes at the nearest available location of Southport (approximately 3 km from the Focus Building) were used for tidal correction. An intertidal beach face slope of 0.06, which is the mean beach slope between 0 and 2 m AHD at Narrowneck [39] was used for the correction.

	TM	ETM+ (SLC-on)	ETM+ (SLC-off)
Period	1999–2008	1999–2003	2003–2008
Number of scenes	63	53	61

Table 1.
Description of Landsat multispectral data used.

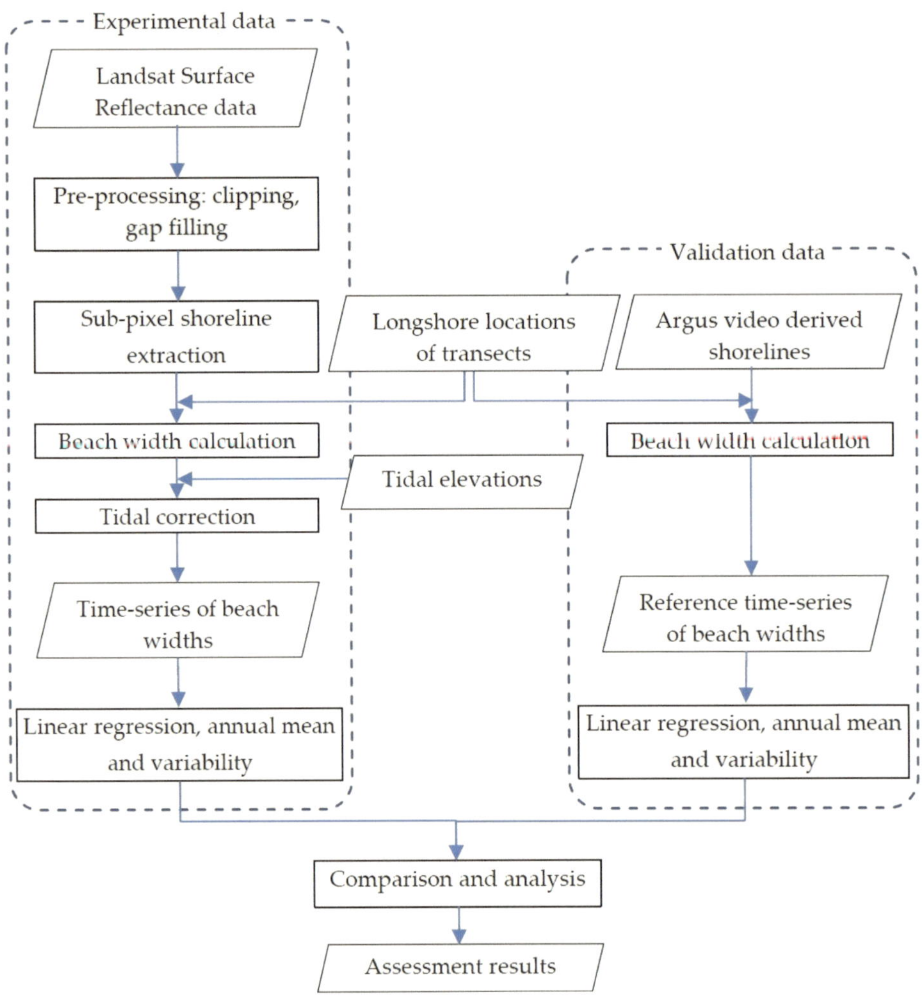

Figure 4.
Flow chart of shoreline mapping and change monitoring.

2.5 Results and discussions

2.5.1 Time-series beach width results

Based on the shorelines derived from Landsat and video imaging data, 100 m longshore-averaged beach widths centred at seven representative transects (marked in **Figure 3(a)**) were calculated. These transects were spaced at a regular interval of 500 m, i.e., they are 2000, 1500, 1000, 500 m to the north, and 500, 1000, 1500 m to the south of the Argus cameras. The selected alongshore locations are consistent with [35] as the report indicated that beach widths at these transects are suitable for the analysis of shoreline trends and variabilities. Using longshore averaged beach widths aims to average out the effects of the longshore varied instantaneous waves at the beach face. The area nearest to the cameras (0 m alongshore) was not selected because of sun glint and the gap between the field views of the cameras [35].

Figure 5 shows the time-series beach widths from 1999 to 2008 at the seven transect lines after tidal correction, compared with video imaging derived results (as reference data). The legend for the alongshore locations were abbreviated according to whether they were north (N) or south (S) followed by their distances, relative to the Argus cameras, e.g. 2000 m at the north was abbreviated as N 2000 and 1500 m at the south as S 1500. The time-series beach widths at the seven locations show very similar trends in the results from the Landsat data compared with the reference data, despite some intermittent results with noticeable errors. The beach widths over the 9 years along the full 4.5 km study area can be seen to have varied up to approximately 100 m.

To realise pairwise comparisons between Landsat and video imaging derived beach widths, each of the final ground truth beach widths was calculated using a linear interpolation between the closest pair of pre and post surveys corresponding to each satellite image acquisition time. A time-series of errors was derived at each transect location as the difference between each pair of Landsat and video imaging derived beach widths. Subsequently, the mean error (ME), mean absolute error (MAE) and root mean squared error (RMSE) of the time-series results from Landsat were calculated as listed in **Table 2**.

2.5.2 Annual mean and variability

Annual mean beach widths from 2000 to 2008 at the seven profile lines were calculated for both Landsat and Argus video imaging based results shown as **Figure 6**. Note that year 1999 was excluded since the available video-based data for that year spanned less than half a year. The inter-annual shoreline change trends derived from Landsat match very well with those from video imaging data, with the most noticeable discrepancies at the northern part during 2000–2003. It was found that during those years the average frequency of the ground survey data and/or the Landsat data at the northern part of the beach (N 2000–N 500) was much lower than average. Taking year 2000 as an example, the average frequencies of the ground survey and Landsat data at the northern part were approximately 17 and 35 days respectively. This is believed to be the main reason for the significant divergence between the annual means of Landsat and ground survey data during that period, as it is believed that higher-frequency data would lead to more accurate estimations of annual mean beach widths.

Table 3 presents the statistical assessment of annual mean beach widths at the seven locations, where the RMSEs are in the range of 3.9–7.2 m. Statistical t-tests of the beach width results from Landsat and video imaging within each year of 2000–2008 were employed, there being nine tests for each of the seven locations. The majority of the tests

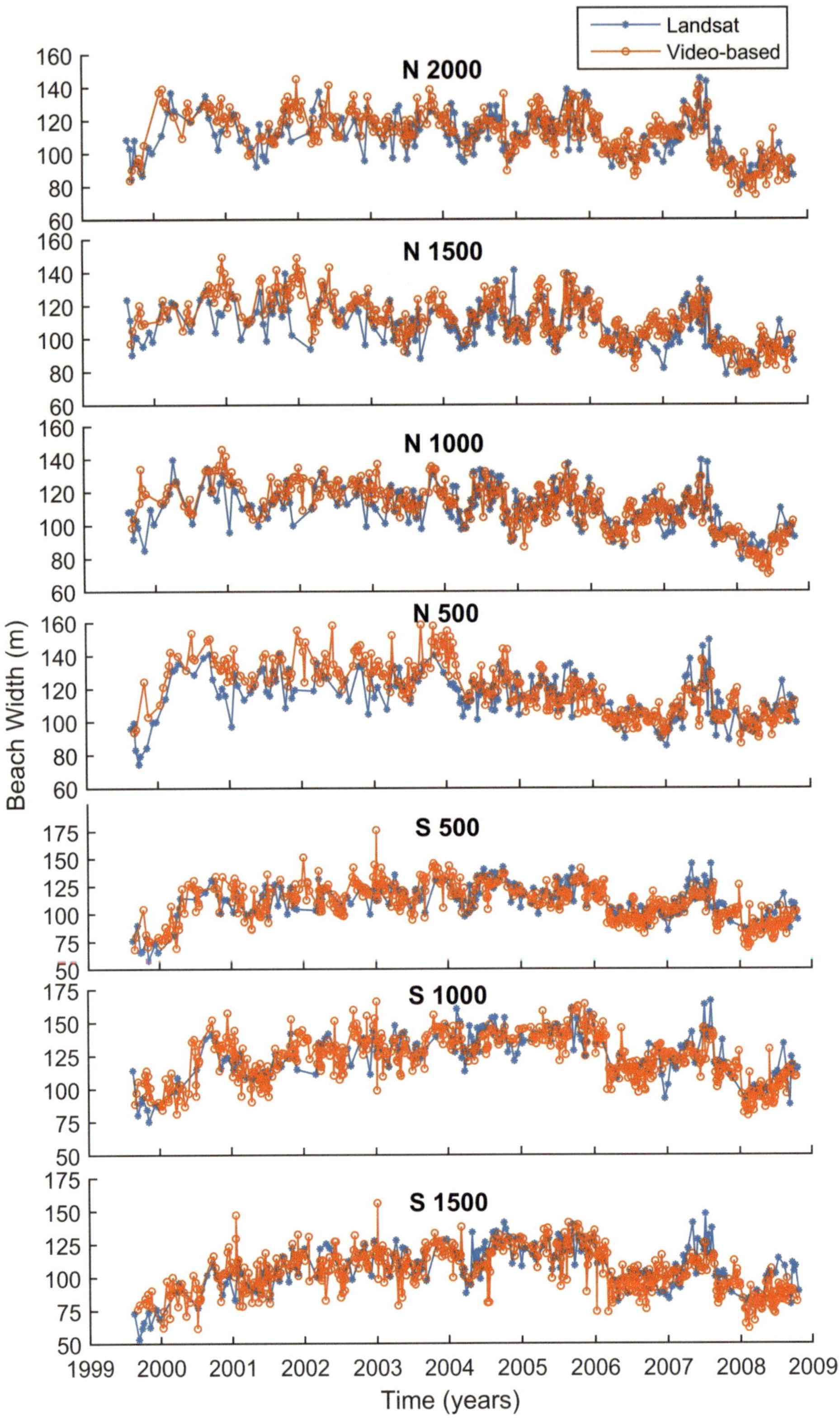

Figure 5.
Time-series beach widths along the seven transects at northern Gold Coast from August 1999 to October 2008. Blue and orange curves represent beach widths from Landsat and Argus coastal video imaging data respectively.

	N 2000	N 1500	N 1000	N 500	S 500	S 1000	S 1500
ME (m)	−1.5	−3.2	−1.0	−4.0	0.0	0.1	1.6
MAE (m)	7.7	8.0	7.2	8.9	9.1	8.8	9.7
RMSE (m)	9.6	10.2	9.1	11.1	11.7	11.4	12.3
Correlation	0.71	0.71	0.72	0.70	0.73	0.78	0.73

Table 2.
Statistical assessment of full time-series beach widths at the seven profile locations.

did not reject the null hypothesis of no significant difference at 5% significance level where only 16 out of the 63 tests rejected the null hypothesis. This indicates the suitability of Landsat data for monitoring annual mean shoreline behaviour at the study site.

The standard deviations within each of the 9 years for the seven locations were calculated and displayed in **Figure 7**. Intra-annual variabilities derived from Landsat data do not appear consistently larger or smaller than the video-based measurements at the northern Gold Coast. The F-test was employed within each of the 9 years to statistically compare beach width variances derived from Landsat and video imaging data. The great majority of tests did not reject the null difference hypothesis of no differences at the confidence level of 95%, where only 6 out of the 63 tests rejected the null hypothesis. This indicates that Landsat data can produce consistent estimations of annual variances with the reference data at the study area.

2.5.3 Shoreline change trends

2.5.3.1 Effects of beach nourishment

It is believed worthwhile to explore the capability of Landsat data to identify the beach width changes predominantly affected by the beach nourishment that occurred between 1999 and 2000. Observing the beach widths (**Figure 8**) from both Landsat and video imaging data during the first year of monitoring (when the nourishment was ongoing), it is clear that the beach widths at the northern part started to show an increase trend soon after the commencement of Argus monitoring project, which is especially observable at N 2000 and N 500. In contrast, the beach widths at the southern part did not show increasing trends until early 2000 when the nourishment at A4–A6 was started. In other words, the time lag of beach widening between the northern and southern parts of the study area as the beach nourishment progressed southwards is observable.

Linear regressions were applied separately to the time-series beach widths from both Landsat and video imaging data from August 1999 to July 2000 and afterwards. The linearly fitted lines are superimposed on the plotted original time-series beach widths in **Figure 8**. The fitted lines at both north and south of the study area show clear increasing trends during August 1999 to July 2000. The regressed changing rates are listed in **Table 4**, which are all positive except for the change rate calculated from video imaging data at N 1000.

Since single beach width results from Landsat data may be erroneous and the time frequency of the extracted shorelines is relatively low, Landsat-estimated linear changing rates during such short periods (i.e. only approximately 1 year) are not reliable, as indicated by the significant differences of the regression rates between Landsat and video imaging results at most of the seven locations. Nevertheless, the most significant accretion rates are at N 2000, N 500 and S 500 for both Landsat and video imaging results. This is consistent with the progress of the nourishment (**Figure 3(b)**) since additional sand was deposited in A1 and A3 (where transects

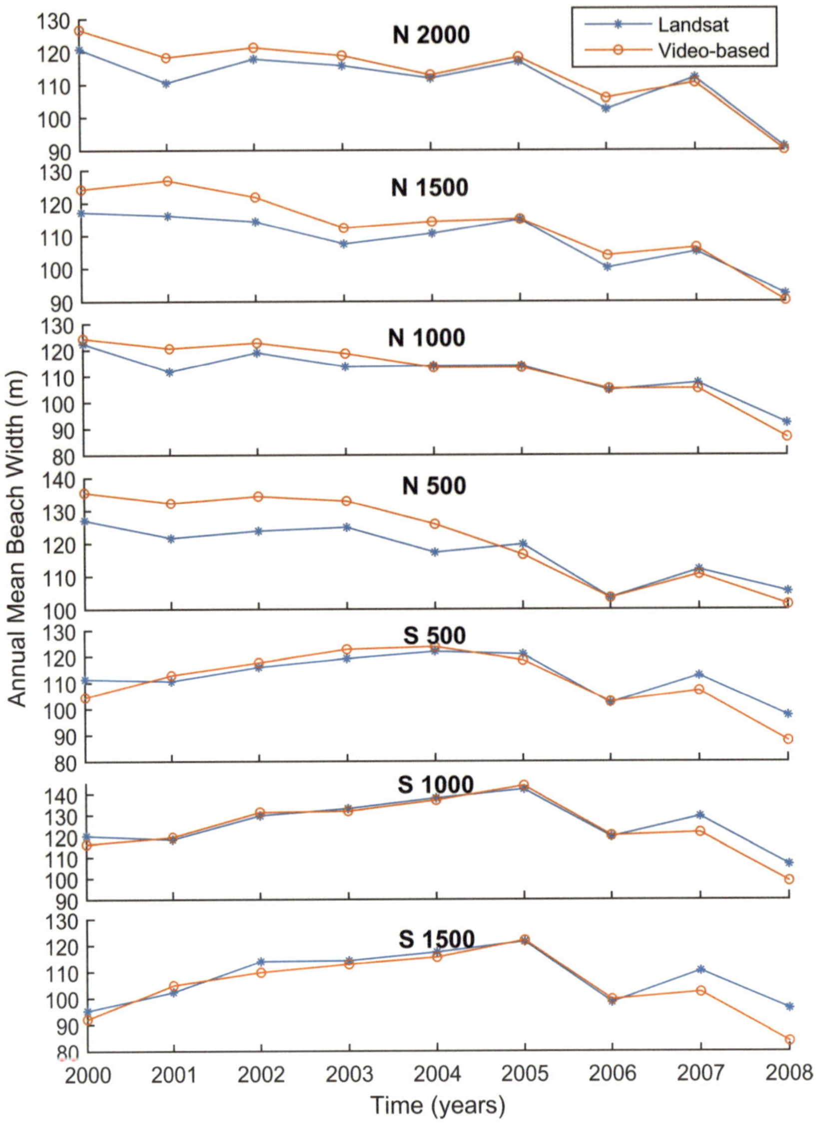

Figure 6.
Annual mean beach widths at the seven locations at northern Gold Coast. Blue and orange curves represent annual mean beach widths from Landsat and from video imaging data respectively.

	N 2000	N 1500	N 1000	N 500	S 500	S 1000	S 1500
ME (m)	−2.6	−4.1	−1.3	−4.3	1.6	1.9	2.9
MAE (m)	3.2	4.6	3.2	6.1	3.8	3.1	3.9
RMSE (m)	3.9	5.6	4.2	7.2	4.7	4.2	5.3

Table 3.
Statistical assessment of annual mean beach widths at the seven locations.

N 2000 and N 500 were located) during the first year of the coastal monitoring, although a large volume of sand had already been deposited before the monitoring commenced. Besides, the sand nourishment at A5, where transect S 500 was located, was implemented after the commencement of the Argus monitoring. In contrast, the nourishment of A2, where transects N 1500 and N 1000 are located,

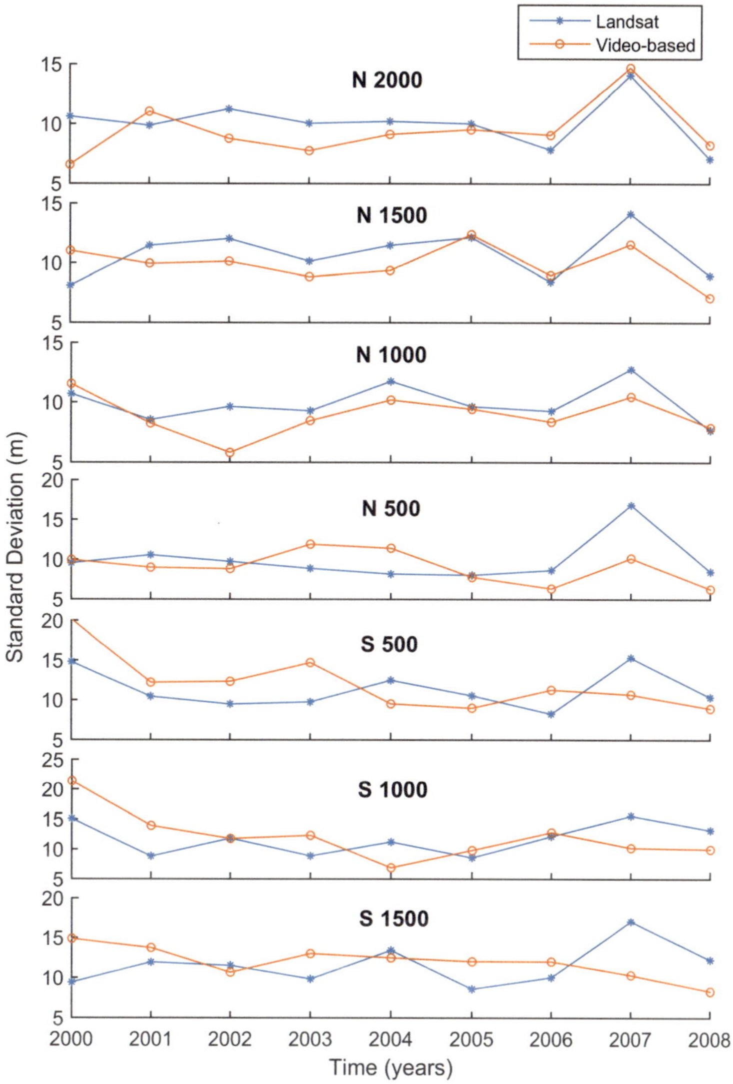

Figure 7.
Annual standard deviations of beach widths at the seven locations at northern Gold Coast. Blue and orange curved represent annual standard deviations from Landsat and coastal video imaging data respectively.

had been completed several months before the coastal monitoring program, and the beach widths over that area are expected to have increased before August 1999. The linear regression results indicate that the effects of beach nourishment progress can be statistically identified from Landsat data.

2.5.3.2 Shoreline change trends post beach nourishment

To quantify the overall trends post beach nourishment at the seven locations, linear regressions were also applied to the time-series results from Landsat and reference data from August 2000 onwards. Slightly decreasing trends were observed at all locations (see the regression lines in **Figure 8**) and corresponding negative changing rates (**Table 4**) were derived from both Landsat and reference data, indicating overall modest erosion trends post the beach nourishment.

Observing **Figure 9** and based on the analysis of shoreline changes illustrated in [35], the post beach nourishment period was judiciously divided into four shorter

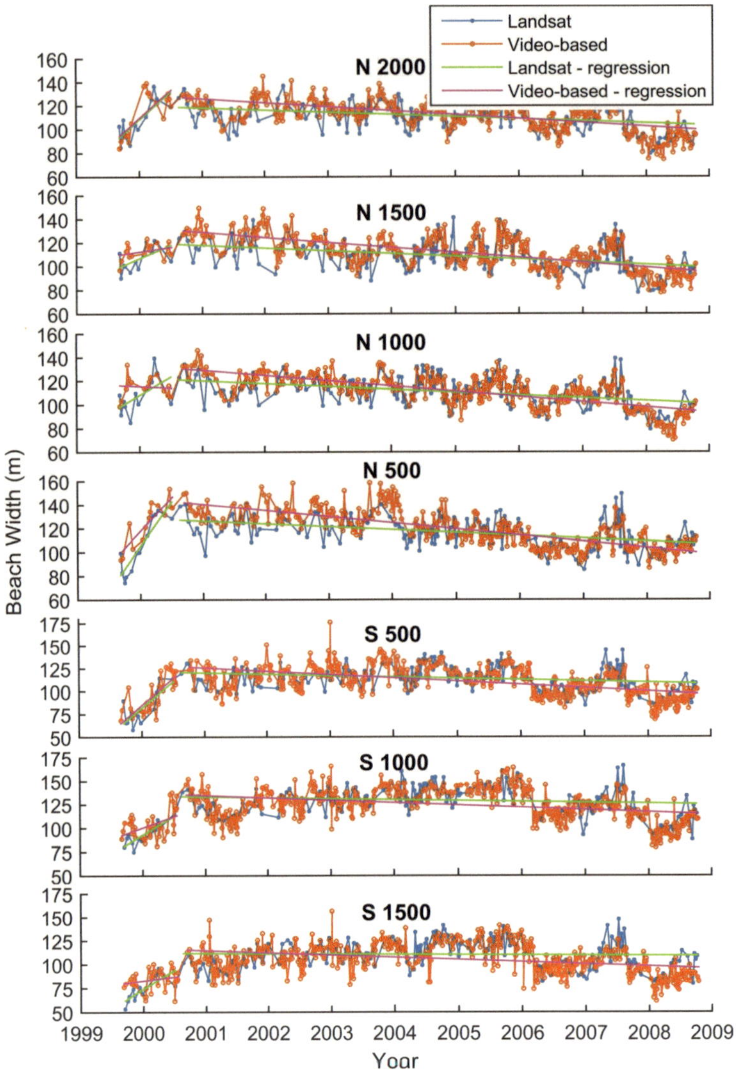

Figure 8.
Linear regression lines of beach widths at the seven profile locations during and after the beach nourishment, where the original time-series of beach widths from Landsat (blue curves) and video imaging data (orange curves) are also plotted for reference.

		N 2000	N 1500	N 1000	N 500	S 500	S 1000	S 1500
Aug 1999–Jul 2000	Landsat	46.6	22.3	30.3	74.4	56.3	39.0	39.4
	Reference	46.9	7.6	−2.5	55.9	58.4	22.7	8.0
Aug 2000–Oct 2008	Landsat	−1.8	−2.3	−2.4	−2.4	−1.5	−1.0	−0.3
	Reference	−3.3	−4.2	−4.4	−5.2	−3.6	−2.5	−2.3

Table 4.
Shoreline changing rates (m/year) of beach widths at the seven locations during and post beach nourishment.

periods: August 2000–February 2006, March 2006–July 2007, August 2007–January 2008 and February 2008–October 2008. Linear regressions were applied to the beach widths during each of the periods and the regression lines are displayed in **Figure 9**.

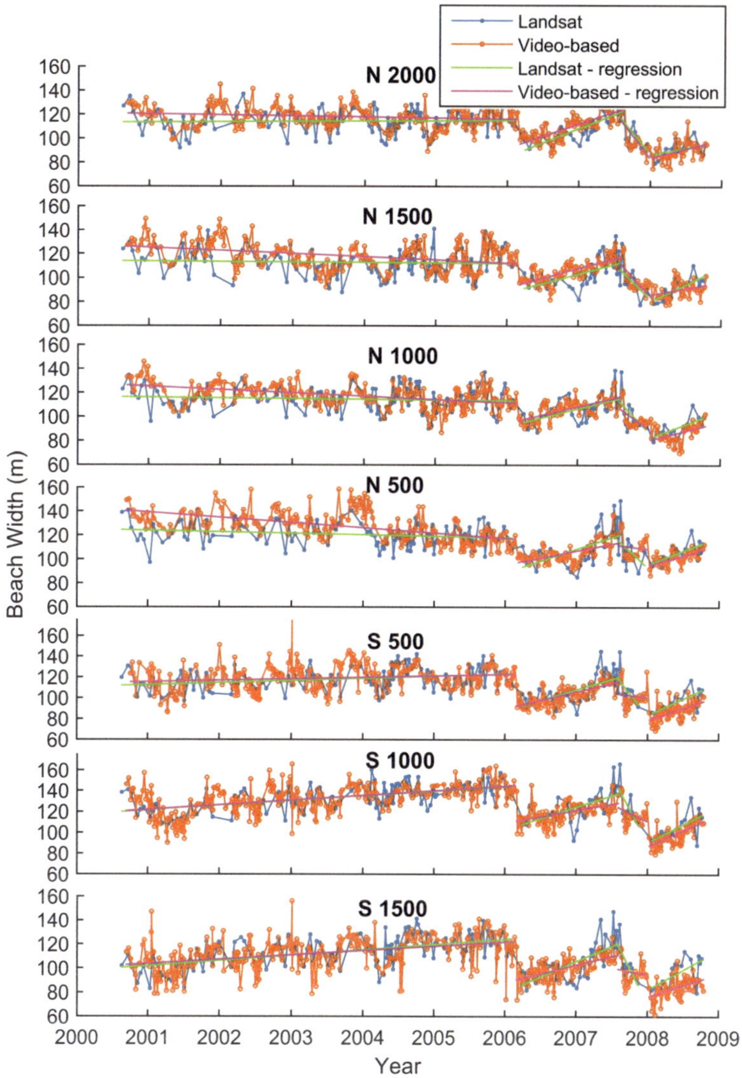

Figure 9.
Linear regressed shoreline change trends during shorter periods post beach nourishment at the seven locations, where the original time-series of beach widths post beach nourishment from Landsat (blue curves) and video imaging data (orange curves) are also plotted for reference.

		N 2000	N 1500	N 1000	N 500	S 500	S 1000	S 1500
Aug 2000–Feb 2006	Landsat	0.28	−0.3	−0.6	−1.2	2.1	4.5	4.4
	Reference	−0.8	−2.6	−2.8	−4.4	1.4	4.5	3.6
Mar 2006–Jul 2007	Landsat	24.8	18.0	15.7	21.8	22.3	20.9	26.1
	Reference	20.6	16.1	14.5	11.2	16.6	11.5	14.5
Aug 2007–Jan 2008	Landsat	−72.7	−77.4	−65.1	−72.7	−121.3	−119.0	−119.5
	Reference	−72.4	−60.4	−39.5	−0.9	−3.8	−25.3	−4.8
Feb 2008–Oct 2008	Landsat	11.0	28.4	22.8	22.9	31.2	34.4	32.9
	Reference	15.2	10.4	15.2	17.1	23.0	30.9	18.7

Table 5.
Shoreline changing rates (m/year) during shorter periods post beach nourishment at the seven locations using linear regression.

The corresponding changing rates during those periods are given in **Table 5**. During the period from August 2000 to February 2006, the northern parts of the study area generally exhibited slight erosions, whereas the southern parts showed modest accretion trend. Dramatic erosions can be clearly observed at all locations in March 2006, as a result of relatively slow passage of an east coast low pressure weather system [35]. In the following months until July 2007, the beach had a steady recovery trend at all locations. During the last period (February to October 2008), the entire beach showed consistent recovery trends.

Note that given such short periods (e.g. less than 6 months), the estimations of net erosion/accretion rates from Landsat are not reliable, as indicated by some significant differences of the regression rates between Landsat and video imaging results. However, the identified trends, i.e. whether accretions or erosions, are still mostly consistent with those from the reference data. This demonstrates the capability of Landsat data to estimate general erosion/accretion trends during periods as short as half a year.

3. Conclusions and recommendation

This chapter presents the application of multispectral satellite imagery for shoreline mapping and change monitoring. Firstly, a brief introduction of shoreline definitions and indicators is given, which is fundamental for shoreline extraction and monitoring. Next, a brief introduction of SRM techniques is presented. A brief review of existing research on time-series shoreline change monitoring based on multi-temporal multispectral satellite imagery is then presented. Most of the studies in the literature used medium to low-resolution satellite data such as Landsat, ASTER and SPOT imagery as the main data source.

Subsequently, a case study of using approximately 9 years of Landsat archival data to monitor shoreline changes at the northern Gold Coast, Australia is presented. By comparing the longshore averaged beach widths at seven representative transects from Landsat with those from Argus imaging data, the derived errors of extracted shorelines are at a sub-pixel level. Specifically, the RMSEs of beach widths at these locations range from 9.1 to 12.3 m and the correlations are all no less than 0.7; the RMSEs of annual mean beach widths are in the range of 3.9–7.2 m. Besides, annual means and variabilities of beach widths can be estimated from Landsat data, without significant differences from the reference data for most of the results. In addition, linear regression results show that Landsat data can be used to identify the general trends of beach widths, i.e. erosion or accretion, during periods as short as half a year. More importantly, beach widening as a result of the sand nourishment can be clearly observed. This case study, together with the existing work in Narrabeen-Collaroy Beach [10], further demonstrates the suitability of Landsat images for long-term shoreline change monitoring and the practicability of the super-resolution shoreline extraction strategy.

On the other hand, Landsat images can be unreliable for the monitoring of shoreline changes during much shorter periods, e.g. only a few days which is the typical duration for a coastal storm event. In this case, satellite images with both higher spatial and temporal resolutions are needed. While Landsat archival data are used in this chapter as a representative data source of medium-resolution satellite data, it is recommended that other multispectral satellite images such as the freely available Sentinel multispectral images could also be investigated.

Conflict of interest

The authors declare no conflict of interest.

Author details

Qingxiang Liu* and John C. Trinder
School of Civil and Environmental Engineering, University of New South Wales, Sydney, Australia

*Address all correspondence to: qingxiang.liu@unswalumni.com

References

[1] Dolan R, Hayden BP, May P, May S. The reliability of shoreline change measurements from aerial photographs. Shore and Beach. 1980;**48**(4):22-29

[2] Boak EH, Turner IL. Shoreline definition and detection: A review. Journal of Coastal Research. 2005;**21**(4):688-703

[3] Harley M, Turner I, Short A, Ranasinghe R. A reevaluation of coastal embayment rotation: The dominance of cross-shore versus alongshore sediment transport processes, Collaroy-Narrabeen Beach, Southeast Australia. Journal of Geophysical Research: Earth Surface. 2011;**116**(F4)

[4] Ruggiero P, Kaminsky GM, Gelfenbaum G. Linking proxy-based and datum-based shorelines on a high-energy coastline: Implications for shoreline change analyses. Journal of Coastal Research. 2003;(38):57-82

[5] Crowell M, Leatherman SP, Buckley MK. Historical shoreline change: Error analysis and mapping accuracy. Journal of Coastal Research. 1991;7(3):839-852

[6] Moore LJ, Ruggiero P, List JH. Comparing mean high water and high water line shorelines: Should proxy-datum offsets be incorporated into shoreline change analysis. Journal of Coastal Research. 2006:**22**(4):894-905

[7] Minogue MJ. Regional-scale coastal response to the April 2015 storm using aerial photography [Honours thesis]. The University of New South Wales; 2016

[8] Aarninkhof SG, Turner IL, Dronkers TD, Caljouw M, Nipius L. A video-based technique for mapping intertidal beach bathymetry. Coastal Engineering. 2003;**49**(4):275-289

[9] Pardo-Pascual JE, Almonacid-Caballer J, Ruiz LA, Palomar-Vázquez J. Automatic extraction of shorelines from Landsat TM and ETM+ multi-temporal images with subpixel precision. Remote Sensing of Environment. 2012;**123**:1-11

[10] Liu Q, Trinder J, Turner IL. Automatic super-resolution shoreline change monitoring using Landsat archival data: A case study at Narrabeen-Collaroy Beach, Australia. Alberta Personal Property Registry Electronic System. 2017;**11**(1):016036

[11] Mertens K. Towards Sub-Pixel Mapping: Design and Comparison of Techniques. [S.l.: s.n.]. Ghent, Belgium: Ghent University; 2008

[12] Mertens KC, de Baets B, Verbeke LPC, de Wulf RR. A sub-pixel mapping algorithm based on sub-pixel/pixel spatial attraction models. International Journal of Remote Sensing. 2006;**27**(15):3293-3310

[13] Mertens KC, Verbeke LPC, Westra T, De Wulf RR. Sub-pixel mapping and sub-pixel sharpening using neural network predicted wavelet coefficients. Remote Sensing of Environment. 2004;**91**(2):225-236

[14] Tatem AJ, Lewis HG, Atkinson PM, Nixon MS. Super-resolution land cover pattern prediction using a Hopfield neural network. Remote Sensing of Environment. 2002;**79**(1):1-14

[15] Liu Q, Trinder J, Turner I. A comparison of sub-pixel mapping methods for coastal areas. ISPRS Annals of the Photogrammetry, Remote Sensing and Spatial Information Sciences. 2016;**III-7**:67-74

[16] Foody GM, Muslim AM, Atkinson PM, editors. Super-resolution mapping of the shoreline through soft classification analyses. In: IGARSS 2003 2003 IEEE International Geoscience and Remote Sensing Symposium Proceedings (IEEE Cat No03CH37477). Toulouse, France: IEEE. 2003;**6**:3429-3431

[17] Foody GM. The role of soft classification techniques in the refinement of estimates of ground control point location. Photogrammetric Engineering and Remote Sensing. 2002;**68**(9):897-904

[18] Atkinson PM. Super-resolution land cover classification using the two-point histogram. In: Sanchez-Vila X, Carrera J, Gómez-Hernández JJ, editors. geoENV IV—Geostatistics for Environmental Applications: Proceedings of the Fourth European Conference on Geostatistics for Environmental Applications Held in Barcelona, Spain, November 27-29, 2002. Dordrecht, The Netherlands: Springer; 2004. pp. 15-28

[19] Muslim AM, Foody GM, Atkinson PM. Localized soft classification for super-resolution mapping of the shoreline. International Journal of Remote Sensing. 2006;**27**(11):2271-2285

[20] Shah CA. Automated Lake shoreline mapping at subpixel accuracy. IEEE Geoscience and Remote Sensing Letters. 2011;**8**(6):1125-1129

[21] Addo KA, Jayson-Quashigah P, Kufogbe K. Quantitative analysis of shoreline change using medium resolution satellite imagery in Keta, Ghana. Marine Science. 2011;**1**(1):1-9

[22] Chen W-W, Chang H-K. Estimation of shoreline position and change from satellite images considering tidal variation. Estuarine, Coastal and Shelf Science. 2009;**84**(1):54-60

[23] Maiti S, Bhattacharya AK. Shoreline change analysis and its application to prediction: A remote sensing and statistics based approach. Marine Geology. 2009;**257**(1):11-23

[24] Harley MD, Turner IL, Short AD, Ranasinghe R. Assessment and integration of conventional, RTK-GPS and image-derived beach survey methods for daily to decadal coastal monitoring. Coastal Engineering. 2011;**58**(2):194-205

[25] Turner IL, Harley MD, Short AD, Simmons JA, Bracs MA, Phillips MS, et al. A multi-decade dataset of monthly beach profile surveys and inshore wave forcing at Narrabeen, Australia. Scientific Data. 2016;**3**. Article number: 160024

[26] Chu Z, Yang X, Feng X, Fan D, Li Y, Shen X, et al. Temporal and spatial changes in coastline movement of the Yangtze delta during 1974-2010. Journal of Asian Earth Sciences. 2013;**66**:166-174

[27] Liu Y, Huang H, Qiu Z, Fan J. Detecting coastline change from satellite images based on beach slope estimation in a tidal flat. International Journal of Applied Earth Observation and Geoinformation. 2013;**23**:165-176

[28] Yu K, Hu C, Muller-Karger FE, Lu D, Soto I. Shoreline changes in west-Central Florida between 1987 and 2008 from Landsat observations. International Journal of Remote Sensing. 2011;**32**(23):8299-8313

[29] Kuleli T, Guneroglu A, Karsli F, Dihkan M. Automatic detection of shoreline change on coastal Ramsar wetlands of Turkey. Ocean Engineering. 2011;**38**(10):1141-1149

[30] Almonacid-Caballer J, Sánchez-García E, Pardo-Pascual JE, Balaguer-Beser AA, Palomar-Vázquez J. Evaluation of annual mean

shoreline position deduced from Landsat imagery as a mid-term coastal evolution indicator. Marine Geology. 2016;**372**:79-88

[31] Li W, Gong P. Continuous monitoring of coastline dynamics in western Florida with a 30-year time series of Landsat imagery. Remote Sensing of Environment. 2016;**179**:196-209

[32] Davidson MA, Turner IL. A behavioral template beach profile model for predicting seasonal to interannual shoreline evolution. Journal of Geophysical Research: Earth Surface. 2009;**114**(F1)

[33] Ruessink BG, Coco G, Ranasinghe R, Turner IL. Coupled and noncoupled behavior of three-dimensional morphological patterns in a double sandbar system. Journal of Geophysical Research: Oceans. 2007;**112**(C7)

[34] Boak L, McGrath J, Jackson A. IENCE—A case study—The Northern Gold Coast Beach protection strategy. Coastal Engineering. Proceedings of the 27th International Conference on Coastal Engineering. 2000;**4**:3710-3717

[35] Blacka MJ, Anderson DJ. Lopez LM. Analysis of shoreline variability, seasonality and erosion/accretion trends: February–July 2008. Report 18. Northern Gold Coast coastal imaging system. WRL Technical Report. November 2008

[36] Aarninkhof S, Roelvink J, editors. Argus-Based Monitoring of Intertidal Beach Morphodynamics. Coastal Sediments. ASCE. Proceedings of Coastal Sediments. New York: American Society of Civil Engineers; 1999;**99**: 2429-2444

[37] Davidson MA, Lewis RP, Turner IL. Forecasting seasonal to multi-year shoreline change. Coastal Engineering. 2010;**57**(6):620-629

[38] Mishra N, Haque M, Leigh L, Aaron D, Helder D, Markham B. Radiometric cross calibration of Landsat 8 operational land imager (OLI) and Landsat 7 enhanced thematic mapper plus (ETM+). Remote Sensing. 2014;**6**(12):12619

[39] Splinter KD, Carley JT, Golshani A, Tomlinson R. A relationship to describe the cumulative impact of storm clusters on beach erosion. Coastal Engineering. 2014;**83**:49-55

Section 3

Infrastructure Investigations, Tsunami Disaster Mapping and Novel 3-D Reconstruction

Chapter 6

Utilization of Dynamic and Static Sensors for Monitoring Infrastructures

Chung C. Fu, Yifan Zhu and Kuang-Yuan Hou

Abstract

Infrastructures, including bridges, tunnels, sewers, and telecommunications, may be exposed to environmental-induced or traffic-induced deformation and vibrations. Some infrastructures, such as bridges and roadside upright structures, may be sensitive to vibration and displacement where several different types of dynamic and static sensors may be used for their measurement of sensitivity to environmental-induced loads, like wind and earthquake, and traffic-induced loads, such as passing trucks. Remote sensing involves either in situ, on-site, or airborne sensing where in situ sensors, such as strain gauges, displacement transducers, velometers, and accelerometers, are considered conventional but more durable and reliable. With data collected by accelerometers, time histories may be obtained, transformed, and then analyzed to determine their modal frequencies and shapes, while with displacement and strain transducers, structural deflections and internal stress distribution may be measured, respectively. Field tests can be used to characterize the dynamic and static properties of the infrastructures and may be further used to show their changes due to damage. Additionally, representative field applications on bridge dynamic testing, seismology, and earthborn/construction vibration are explained. Sensor data can be analyzed to establish the trend and ensure optimal structural health. At the end, five case studies on bridges and industry facilities are demonstrated in this chapter.

Keywords: health monitoring, accelerometers, velometers, displacement transducers, strain sensors, frequency response function, cross-power spectrum, power spectral density, bridge dynamic testing, seismology, earthborn/construction vibration, infrastructure

1. Introduction

In order to acquire infrastructural health data, proper sensor knowledge and technology are required. This article first introduces in situ remote sensing and then provides a review of some sensors that are useful and currently implemented in health monitoring projects, especially those associated with vibration.

A project on the development of a self-sustained wireless integrated structural health monitoring (ISHM) system for highway bridges was sponsored by the USDOT Research and Innovative Technology Administration (RITA) [1]. **Figure 1** shows the wireless ISHM system with remote sensing ability: (1) wireless sensor nodes including AE sensors, strain gages, accelerometers, thermocouples, etc.; (2) wireless smart sensor network with (3) energy harvester; (4) data acquisition system (DAQ), with wireless communication modem; and (5) web-based remote data processing and data storage for application.

In situ sensors may include the capability to collect static and dynamic data and then apply algorithm to extract and combine relevant condition information from sensor data. Typical vibrational sensors used include accelerometers and velometers (velocity transducers), while static sensors include displacement transducers, strain gauges (transducers), tilt meters, and weather-related sensors to measure and record temperature, humidity, barometric pressure, wind velocity, wind direction, etc. When using vibration data, especially in conjunction with modeling systems, the data is often measured in the form of acceleration, velocity, and displacement. Sometimes different analyses require measured signals in different forms. Even if we measure in the form of acceleration, velocity, or displacement (**Figure 2**), we may apply simple mathematics to convert between them through integration or differentiation. For instance, if the measured signal is from accelerometers, we may obtain the velocity through integration and displacement through double integration. On the other hand, if the measured signal is from velocity or displacement

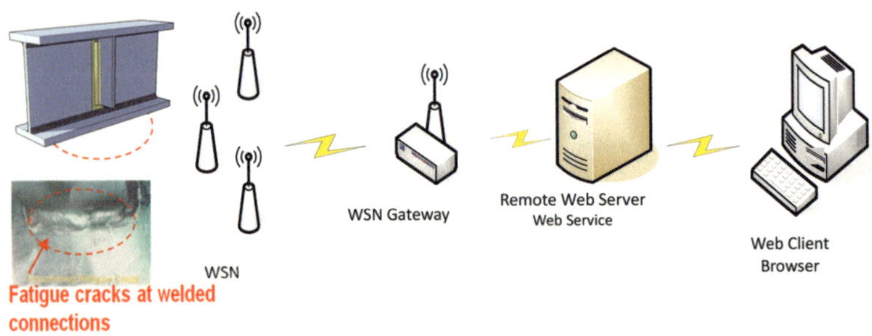

Figure 1.
Remote wireless bridge monitoring system.

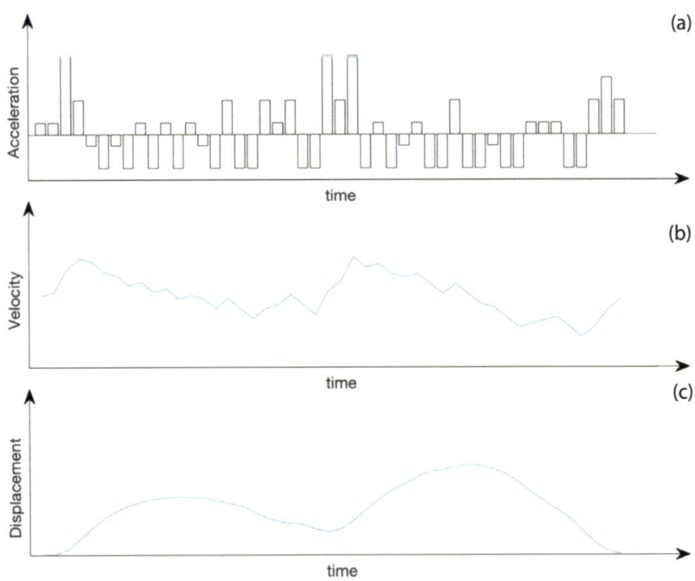

Figure 2.
Measured signals in different forms: (a) acceleration (raw data), (b) velocity (single integration from acceleration), and (c) displacement (double integration from acceleration).

transducers, we may obtain acceleration through differentiation or double differentiation, respectively. Usually, unless there are special circumstances, the suggested method to measure vibration is with an accelerometer. However, care is required to remove accelerations of very low frequencies for possible noises if any integration to velocity or displacement is needed.

Accelerometers and strain sensors are widely used dynamic and static monitoring sensors. The modern-day systems are small, lightweight, and robust and are typically quite simple to calibrate and to convert output to acceleration or strain data. Accelerometers are useful for measuring with low to very high sampling rates. They have shown to be useful in a wide variety of applications. On the other hand, velocity sensors are generally used to measure dynamic response in the low- to medium-range frequencies. They are typically used for similar applications as accelerometers [2].

For the static monitoring sensor, displacement transducers are used to measure relative displacement. These sensors are available in both contacting devices, like string pot and linear variable differential transformer (LVDT), and non-contacting devices, like laser displacement, global positioning systems (GPS), and photogrammetry. The major limitation for contacting displacement-measuring devices in the field is that the measured displacement is a relative displacement. GPS-type sensors are gradually more often used in civil engineering studies because of recent developments allowing measurements to be taken at high fidelity. Displacement measurements from laser sensors, ultrasonic distance sensors, and strain pot were used on different occasions to determine the vertical deflection of a bridge. These techniques are useful because they can result in relative and absolute displacement states. Strain sensors, including optical fiber strain, can be monitored at dynamic rates, while traditional foil strain gauges have been widely used on civil engineering structures, even in remote sensing.

2. Mathematical models for computing accelerometer sensor data

The data acquisition system may be set to measure acceleration time histories and calculate frequency response function (FRF), cross-power spectrum (CPS), and power spectral density (PSD) [3].

For a continuous time series, $x(t)$, defined on the interval from 0 to T, the Fourier spectrum (Fourier transform), $X(f)$, is defined in Eq. (1) as

$$X(f) = \int_0^T x(t)e^{-i2\pi ft}dt \qquad (1)$$

where $i = \sqrt{-1}$ and $f = cyclic\ frequency(Hz)$.

This function is complex, and the magnitude is typically plotted in engineering units (EU), such as m/s^2 or g's, versus frequency.

This power spectrum is defined in Eq. (2) as

$$|X(f)|^2 = X(f)X(f) \qquad (2)$$

where * denotes a complex conjugate. The power spectrum is a real-valued frequency domain function and has the units of $(EU)^2$.

The power spectral density (auto-spectral density, or abbreviated as PSD), $G_{XX}(f)$, is defined in Eq. (3) as

$$G_{XX}(f) = \frac{2}{T}E\left[(X(f))^2\right] \tag{3}$$

where $E[n]$ indicates an ensemble average for a specific t over n samples of $X(f)$. This PSD is a real-valued frequency domain function and has the units of EU^2/Hz.

The cross-power spectrum (cross-spectrum density, or abbreviated as CPS), $G_{XY}(f)$, relating two time histories, $x(t)$ and $y(t)$, is defined in Eq. (3) as

$$G_{XY}(f) = \frac{2}{T}E[X(f)Y(f)] \tag{4}$$

For a linear system, the frequency response function (transfer function, or abbreviated as FRF), $H(f)$, which relates an input $X(f)$ to a response $Y(f)$, is defined in Eq. (5) as

$$H(f) = \frac{Y(f)}{X(f)} = \frac{G_{XY}(f)}{G_{XX}(f)} \tag{5}$$

In actual dynamic testing, discrete time series are measured. Refer to Bendal and Piersol [4] for the discrete representations on the functions listed in Eqs. (1)–(5).

There are several factors that would affect system-level measurement accuracy, which are (1) sensitivity error and initial absolute offset, (2) nonlinearity of the data, (3) total offset variation from initial absolute offset, and (4) noise. To improve the accuracy, two- or three-point calibrations recommended by manufacturers may be needed.

The output spectrum (measured with accelerometers) can be assumed to be linearly related to the input spectrum through the FRF, which contains both resonant frequency and damping information of the vibrating system. Resonant frequencies can be determined from peaks in the output spectrum, and damping values can be determined by the half-power bandwidth (HPBW) method.

The damping ratio, or damping coefficient, ξ, is defined as $c/c_c = c/2\sqrt{km}$ to be used in the dynamic analysis. Normally, steel bridges have a low damping coefficient $\xi \leq 0.02$. The half-power (bandwidth) method is the most commonly used experimental method [5] to determine the damping in the structure by using two frequencies shown in **Figure 3** and Eq. (6):

$$\xi = \frac{f_2 - f_1}{f_2 + f_1} \tag{6}$$

Mathematically the most common and easy way is to use the Rayleigh damping method with a linear combination of the mass and the stiffness matrices as Eq. (7):

$$c = a_0 m - a_1 k \tag{7}$$

where c, m, and k are the damping, the mass, and the stiffness matrices, respectively, a_0 and a_1 are proportional constants, and c_c represents the critical damping coefficient. The relationship between the damping ratio and the frequency for Rayleigh damping is shown in **Figure 4**. By simplification, these lead to Eq. (8):

$$\begin{Bmatrix} a_0 \\ a_1 \end{Bmatrix} = \frac{2\xi}{\omega_n + \omega_m}\begin{Bmatrix} \omega_n\omega_m \\ 1 \end{Bmatrix} \tag{8}$$

The CPS plot between the signals from two accelerometers can then be used to determine the vibration mode shape information based on the relative phase of

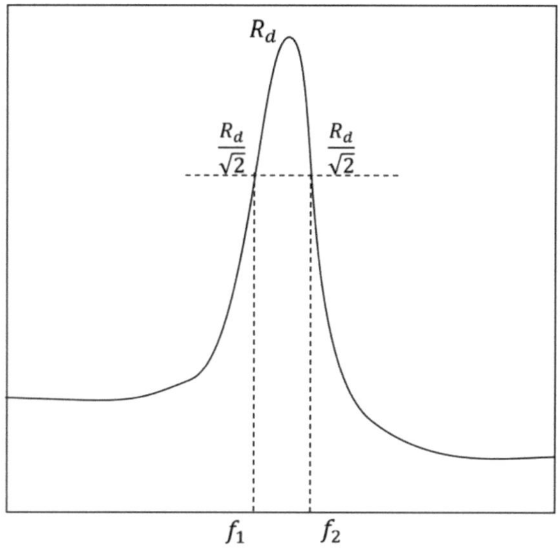

Figure 3.
Half-power method to estimate damping by experiment.

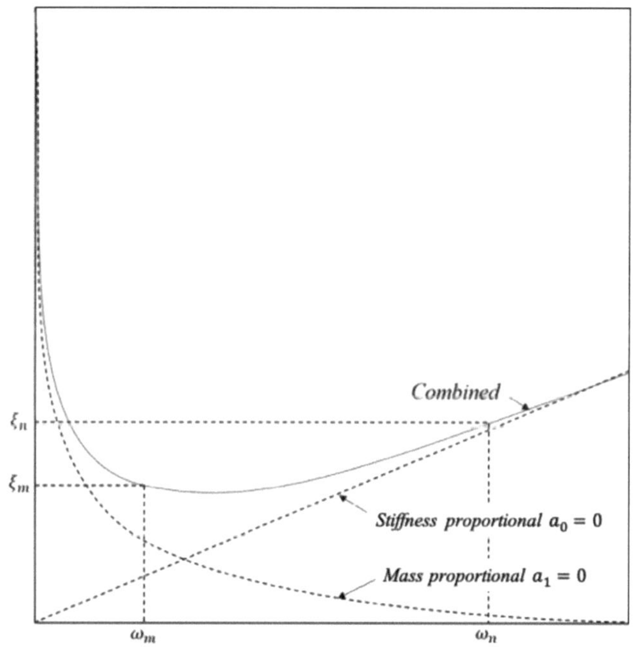

Figure 4.
Relationship between damping ratio and frequency for Rayleigh damping.

the two signals. One signal is termed the reference signal, and the process is repeated at various stations on the bridge to map out the mode shapes. Typically, in vibration testing FRFs are used to estimate the dynamic properties of a structure. Further interpreted from the CPS, it can be seen that two measured responses are correlated only at the resonant frequencies of the structure. Therefore, the CPS will show peaks corresponding to the resonant frequency which shows another method estimating the resonant frequencies from peaks in the response power spectra. Mode shapes are estimated from the relative magnitudes of these peaks, where relative phase information can be obtained from either the CPS or FRF and modal damping values can be obtained by applying the HPBW method to these peaks, which need very-high-frequency resolution to obtain the values. Mode shapes can be determined from cross-power spectra of the various accelerometer readings relative to the reference accelerometer [3]. Examples of field dynamic applications are shown in the next sections.

3. Representative applications

3.1 Bridge dynamic testing

Dynamic testing on bridges has been conducted for many years. Measured data were usually in the form of deflections and strains, but some measurements were in acceleration. For bridge dynamic testing, ambient and forced vibrations can be performed.

- Ambient vibration testing—Ambient vibrations in bridges can be induced by a wide variety of environmental factors, such as traffic, seismic, and wind loading.
- Forced vibration testing—Some techniques of forced vibration testing of bridges such as variable frequency rotating dynamic shaker, servo-hydraulic inertial actuators, impact hammer, and controlled truck loading can be applied. Accelerometers can be used to determine the resonant frequencies, damping ratios, and mode shapes.

By using accelerometers, acceleration time histories can be obtained, transformed into Fourier spectra and CPS, and then analyzed to determine damping, resonant frequencies, and corresponding modal shapes.

3.2 Seismology

Devices can be used to measure seismic data. Two types of sensors (transducers) were used by Caltrans to measure seismic record [6].

- Seismometer—A seismometer, also called a velocity transducer, measures velocity directly using a signal conditioner. It measures low frequencies of ground motions (usually 1–200 Hz) and produces a voltage proportional to velocity through magnetic induction. A seismometer can catch low rate vibrations during monitoring.
- Accelerometer—An accelerometer measures acceleration directly by using the piezoelectric crystal material. This type of sensor, which is widely used by Caltrans, is pressure sensitive and can also obtain velocity and displacement with an integrator. Accelerometer is usually a small sensor with a wide frequency range, and typically not as sensitive as the seismometer. The frequency range could be narrowed from 0.1 to 1.0 KHz when using as large sensor as around 1 pound in weight and more sensitive technical methods, typically from 1.0 to several KHz.

3.3 Earthborn/construction vibration

Humans have varying sensitivities to vibrations at different frequencies. In general, humans are more sensitive to low-frequency vibration. Construction activities could induce vibrations that caused building surface movements, shaking or rattling of windows, hanging items, and lightweight furniture [7]. This type of low-frequency vibrations, when acting on the structural component, can also produce an audible rumbling noise, which referred to earthborn noise. The noise could be a problem when the upper end of the range frequencies (60–200 Hz) dominates the originating vibration spectrum, or the construction activities are connected to the structure by foundations or utilities.

Earthborn vibrations can be detected and measured by accelerometers which could be mounted to heavy blocks of steel (about 5–10 kg) directly placed directly on the ground or other surfaces by magnets [6, 8]. Activities and motions of the vibration-sensitive land shall be monitored and measured during constructions occur within 15 m (50 ft) to establish the level of vibrations. Construction projects of foundations, like pile driving, jackhammering, and soil compacting, may also produce high-level vibrations by their equipment operations. Measured vibration data from construction are commonly classified as broadband or random vibrations with various ranges of frequencies. The general frequency ranges of most earthborn vibrations are from less than 1.0 to 200 Hz.

Vibration levels can be represented in terms of velocity (in/sec or mm/sec) or acceleration (in/sec^2 or mm/sec^2), which demonstrates vibration severity. Vibration levels for construction activities are recognized as the highest during demolition activities and soil compacting. Vibration levels are required to remain below 0.5 in/sec (15 mm/sec) at residences along the project corridor and minimized risk for structural damage. Vibration levels from other general construction activities will also be well below the 0.5 in/sec (15 mm/sec) criteria.

The US Department of Transportation (USDOT) has guidelines for vibration levels from construction related to their activities and recommends that the maximum peak-particle-velocity levels remain below 0.05 in/sec (1.5 mm/sec) at the nearest structures. Vibration levels above 0.5 in/sec (1.5 mm/sec) have the potential to cause architectural damage to normal dwellings. The USDOT also states that vibration levels above 0.015 in/sec (0.45 mm/sec) are sometimes perceptible to people and the level at which vibration becomes annoying to people is 0.64 in/sec (19.2 mm/sec).

3.4 Types of accelerometers and their advantages/disadvantages

Popular types of accelerometers used in the infrastructural areas are (1) bulk micromachined capacitive, (2) bulk micromachined piezoelectric resistive, (3) capacitive spring-mass system based, and (4) laser accelerometers [9].

The work principles of different types of accelerometers are based on piezoelectric effect due to accelerative forces and displacement sensing based on displacement of mass. The advantages of piezoelectric resistive are (1) rugged and inexpensive, (2) high impedance, (3) high sensitivity, and (4) high-frequency response. However, their disadvantages are (1) sensitive to temperature, (2) hysteresis error, (3) less longevity, and (4) decreased efficiency with time.

On the other hand, displacement sensing or seismic-type accelerometers are using spring-mass-damper system, and their advantages are (1) easy calculation, (2) simple and reliable, and (3) durable and efficient. Their disadvantages are (1) spring system not always accurate and (2) fluctuation in mass leading to wrong calculation.

4. Case study 1: Wireless accelerometer sensing of a self-sustained wireless integrated structural health monitoring (ISHM) system on Beaufort #25 bridge, NC

A scalable integral structural health monitoring (ISHM) system sponsored by the USDOT had been developed by the University of Maryland (UMD) and North Carolina State University (NCSU) with the URS (later named AECOM) Corporation [1]. This system, with remote sensing capability, is designed to be suited for fatigue condition assessment of highway steel bridges. Furthermore, the ISHM system would help in damage detection and deterioration diagnosis in early stages, predicting the remaining service life more accurately when compared with the traditional SHM system with reliable technology to improve current inspection methods, and reduce the operating and maintenance costs.

The ISHM system based on wireless sensor networks entails a few recent innovations which applied the current state of the practice in remote sensing and highway infrastructure management. Accelerometers, in this system, are used for monitoring the vibration response of bridges so that the modal frequency information could be obtained and used to calibrate the finite element model of the monitored bridge.

In this system, a new wireless piezoelectric sensor board had been designed and used. This board mainly consists of an 8-bit microcontroller, a FPGA, and a piezoelectric amplifier circuit. This device is enhanced with improved operating frequency and a four-wire, SPI-compatible interface while having lower power consumption. In the ISHM system, each single wireless sensor was tested on a shaker to verify that the developed sensor can recover the input information accurately. However, a single sensor could not catch enough data for structure monitoring and analysis. Thus, a number of wireless sensors along the bridge span are needed [1].

The example for the ISHM accelerometer monitoring case is the Structure No. 060025 Swing Bridge in Beaufort County, North Carolina (**Figure 5**, Beaufort #25 Bridge). The bridge consists of side spans and main spans. It should be noted that the structural support of the side span is a simply supported steel girder bridge, which has a relatively simple stress state compared with the main span because the

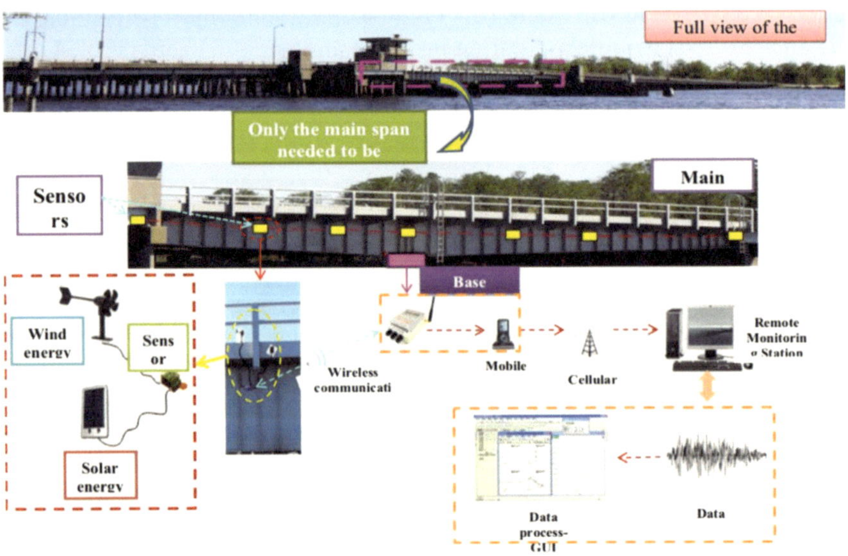

Figure 5.
Sketch plan for monitoring system.

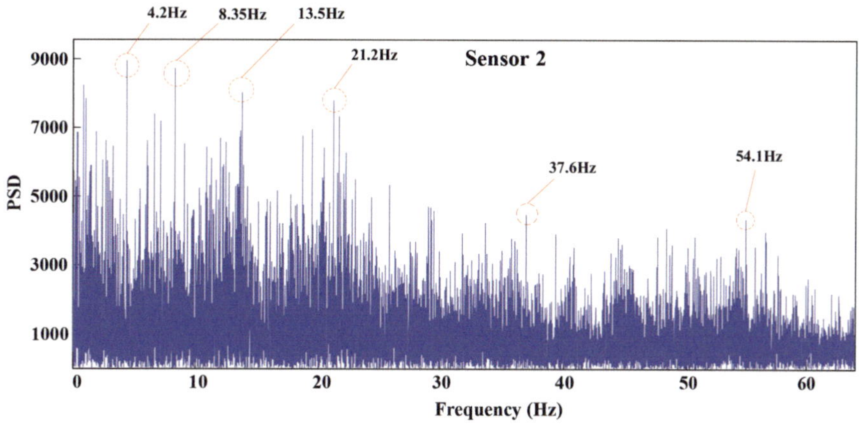

Figure 6.
The result of the field test of Beaufort #25 bridge.

boundary condition of the main span is changed between simply supported and cantilever due to the close or open of the main span. Thus, the researchers of NCSU chose the main span as the targeted monitoring case for the dynamic behavior considering the complex stress states.

In this case, a row of smart sensors was attached to the bridge girders in the main span. The dynamic behavior was analyzed by data from accelerometers. **Figure 6**

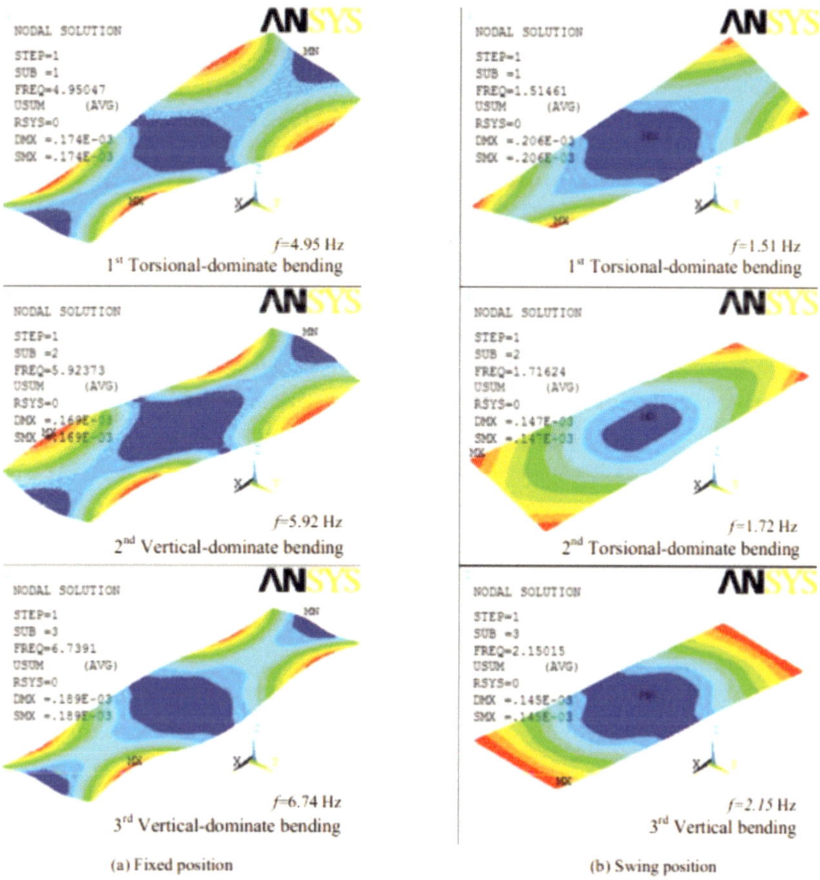

Figure 7.
The first three mode shapes from FE analysis.

Fixed position		Swing position	
First (torsional)	4.95 Hz	First (torsional)	1.51 Hz
Second (vertical)	5.92 Hz	Second (torsional)	1.72 Hz
Third (vertical)	6.74 Hz	Third (vertical)	2.15 Hz
Fourth (vertical)	7.48 Hz	Fourth (vertical)	2.51 Hz
Fifth (lateral)	8.18 Hz	Fifth (vertical)	2.58 Hz

Table 1.
Modal analysis results.

shows the test results of the bridge by using the set of wireless sensors. The data is processed using fast Fourier transform (FFT). The estimation of the natural frequency of the bridge about 4.0 Hz to 5.0 Hz was made by the NCSU researchers.

Meanwhile, the finite element model using the software ANSYS of Beaufort #25 Bridge was built and analyzed. The structural analysis was separated into two conditions due to the fact that the main span could swing. The first three mode shapes are illustrated in **Figure 7**, and the first five modes are summarized in **Table 1**. Depending on the relative amplitude of the mode shapes, these modes were noted as the vertical-dominated modes, the lateral-dominated modes, and the torsional-dominated modes (**Figure 7**).

The accelerometers are commonly used in highway bridges' monitoring for dynamic behavior. The monitoring results for the bridge are close to the finite element analysis result, and thus, the model was calibrated to be analyzed for other load conditions, and the test results were archived to be the baseline for future monitoring.

5. Case study 2: Remote monitoring of a self-sustained wireless integrated structural health monitoring (ISHM) system for highway bridges on I-270 bridge in MD

The second case study is under the same ISHM project [1] and was conducted by the University of Maryland at College Park. The types of sensors used in this project were (1) piezoelectric paint AE sensors; (2) wireless accelerometers; (3) laser sensor; (4) ultrasonic distance sensors; (5) BDI strain transducers; and (6) string pots.

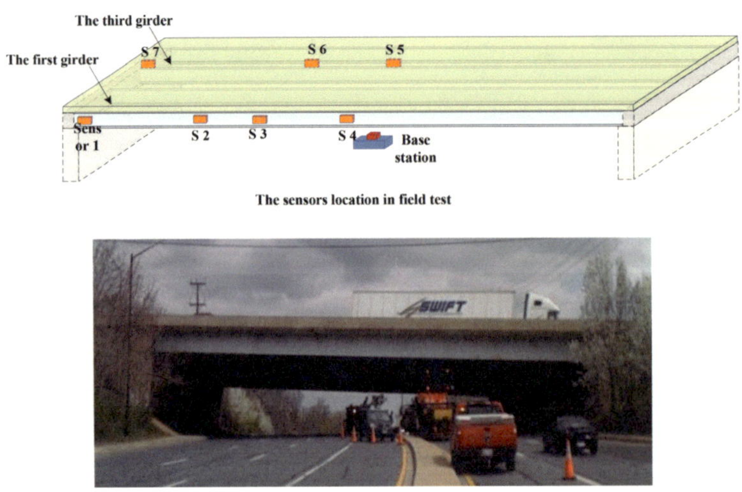

Figure 8.
Sensor locations.

DOI: http://dx.doi.org/10.5772/intechopen.83500

In order to verify the reliability of the whole system, a field test for I-270 Bridge in Maryland by using this ISHM system was carried out with the accelerometer sensor locations shown in **Figure 8**. **Figures 9** and **10** show the test results collected by these wireless sensors.

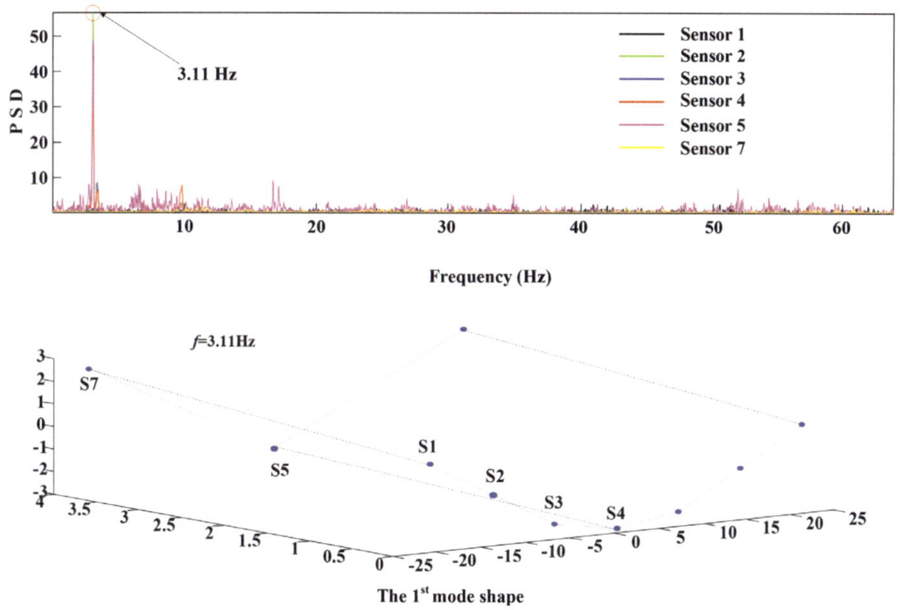

Figure 9.
PSD of these sensors and the first mode shape of the bridge.

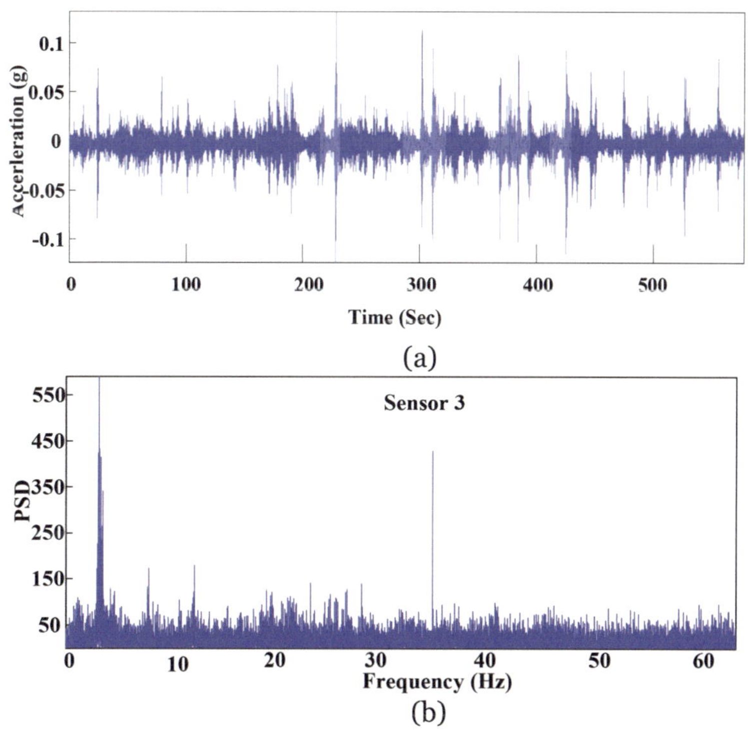

Figure 10.
The results of field test of I-270 bridge, MD: (a) the time-history data of sensor 3 and (b) the PSD of sensor 3.

6. Case study 3: Wireless structural monitoring of a newly replaced fiber-reinforced plastic (FRP) bridge deck

The use of FRP-composite bridge decks is viewed as a potential long-term solution for the concrete deck deterioration problem. A pilot project sponsored by the Federal Highway Administration (FHWA), USA, was undertaken by the Maryland State Department of Transportation, partnered with the University of Maryland to rehabilitate a steel truss bridge (MD24 over Deer Creek in Harford County, Maryland) using lightweight FRP deck [10, 11]. The existing steel truss bridge (**Figure 11**), built in 1934, carries two lanes of traffic, provides 9.14 m (30 ft.) of clear roadway, and is 37.50 m (123 ft.) long with severe roadway skew (**Figure 12**). The FRP deck panels are placed perpendicular to the stringers and act as a continuous plate between the stringer supports.

Load tests and structural monitoring were conducted to obtain information regarding the performance of the structure. For a relatively new material like FRP, the use of load tests can prove the structure's capacity. Wireless structural monitoring system developed through a previous FHWA small business innovation research (SBIR) contract to Invocon, Inc. in Conroe, Texas, was used. The system includes a data acquisition and communication nodes (**Figure 13**) connected to strain gages that can acquire data in digital form and relay the

Figure 11.
Steel truss bridge on MD 24 over deer creek.

Figure 12.
Replacement of a FRP deck panel.

data to a local base receiver attached to a personal computer. In this load test, five boxes were linked in a "smart" network to control the data acquisition process. By using this system, the effort of instrumenting a bridge was reduced by more than half compared to hardwired systems. All CEA-06-250-UN350 uniaxial gages installed on the bridge are produced by the Measurements Group, Inc. As shown in **Figure 13**, strain gauges were strategically placed at different locations to measure strains due to live load effect. Three stringers, as shown in **Figure 13(a)**, were load tested to check the distribution of live load over the stringers. Strain gages (data sets 2–1, 2–2, and 2–3 in **Figure 13(b)**) were located on the top of bottom flanges in the middle of the span. Comparison of finite element results and test results shows that the percentage difference ranged between 1.47 and 9.43%. The purpose of this test is to prove the integrated composite action between the steel stringers and the new FRP panels [10, 11].

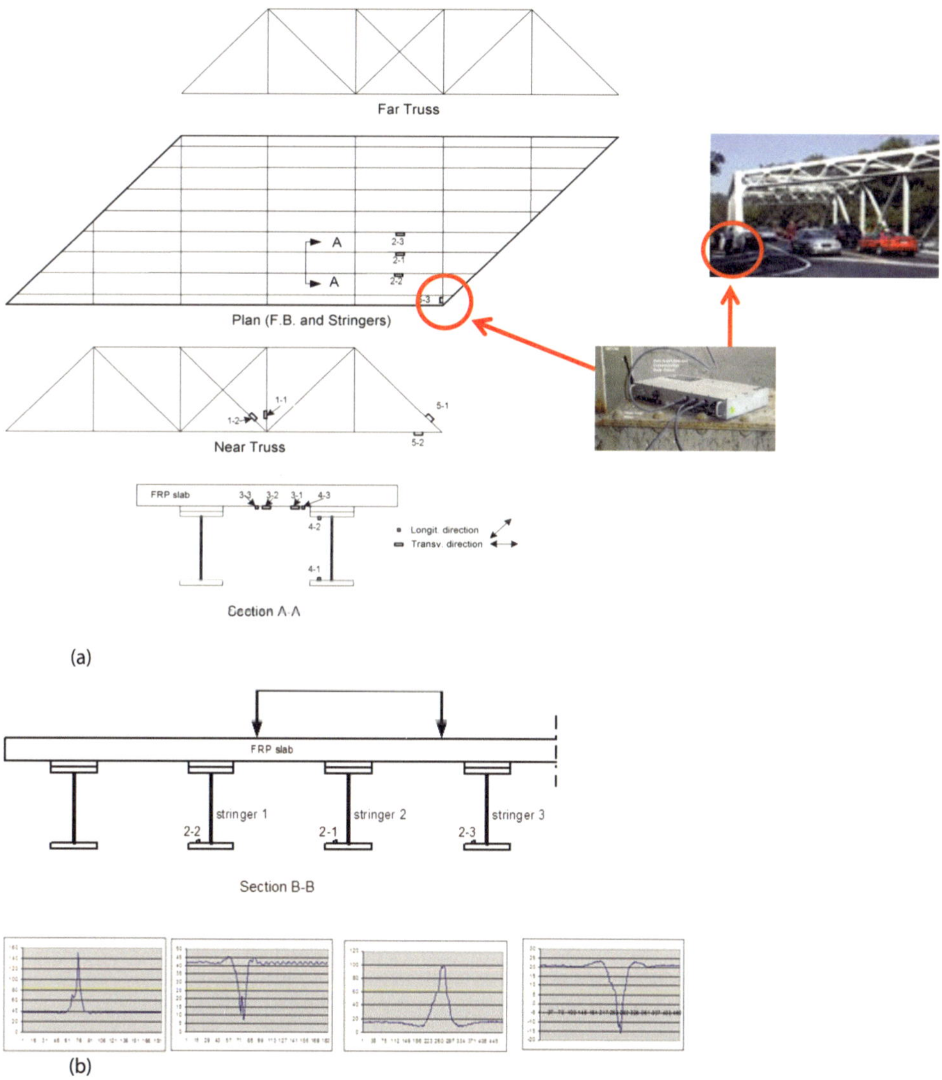

Figure 13.
Truss bridge deck, stringers, strain sensor locations, and data: (a) plan, elevation, and section A-A views and (b) section B-B and strain data measurement.

7. Case study 4: Digital accelerometer monitoring of hanger cables on arch bridges

Arch-girder bridges with hanger cables are a popular type of bridges because they have the advantages of both arch and girder forms. Therefore, it is critical to check the performance of the hanger cables in order to guarantee road safety. The hanger, which ties the arch and the girder, is a key determinant of bridge quality. If one hanger is damaged, the whole structure is at risk. By detecting bridge's hangers, we may make judgment whether the bridge is in good condition or not:

$$T_0 = ml^2\left(4.3865f_1^2 - 0.2742f_2^2\right) \tag{9}$$

where T_0 is the cable tension, m is the mass of the cable, l is the length of the cable, and f_1, f_2 are the first and second natural frequencies, respectively.

In Eq. (9), the stiffness of hanger cable is not needed to be tested, only frequencies. Therefore, it has an advantage of easy operation and usage. The demonstrated bridge here is a tied-arch bridge, and the above equation was used to calculate the cable forces, which are shown in Refs. [5, 12].

In the project, digital accelerometer JMM-268 dynamic testing instrument (**Figure 14**) was used to measure the first and second frequencies of hanger cables. When the frequencies were obtained, the hanger cable force can be calculated according to Eq. (9).

Comparing calculated hanger cable forces with cable force capacity, inspector of the bridge can locate critical sites and focus on those sites to do more detailed inspection. With the fast assessment method presented, only the first and second frequencies of the hanger cable need to be detected. This method was used to evaluate several arch bridges with hanger cables [5, 12].

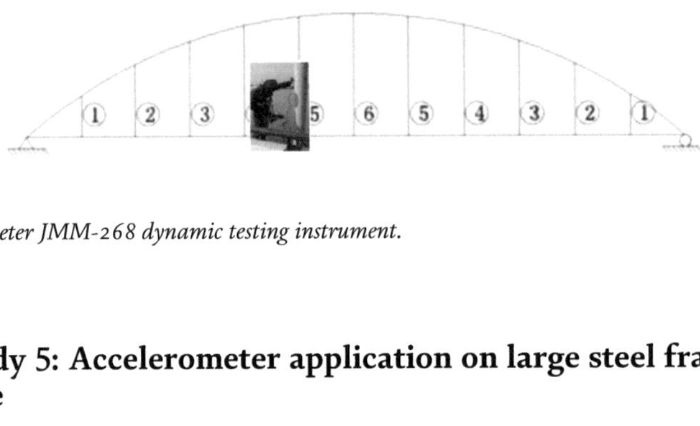

Figure 14.
Digital accelerometer JMM-268 dynamic testing instrument.

8. Case study 5: Accelerometer application on large steel frame structure

The steel frame structure is commonly used in the infrastructure of the petroleum industry to support numerous pipes and storage tanks. Vibration in steel frames is an industrial safety issue due to the movement of massive amounts of liquid, solid, and gas through the pipes. According to statistics published by David G. Maboney [13] regarding the causes of serious disasters in petrochemical industries, tube systems took up to 33% of the equipment. To identify the structural behavior of steel frames, accelerometers could be applied to detect vibrations. Shown here is an industry case that a new half-mile-long, 76-in (190 mm) diameter pipeline system is installed above a large 80-ft (24 m)-high steel frame structure in an oil refinery in Taiwan (**Figure 15**). The main function of the 76-in pipeline is to deliver massive amount of waste to the flare stack. However, an unexpected disturbance of the 76-in pipeline occurs which becomes the source of dramatic and

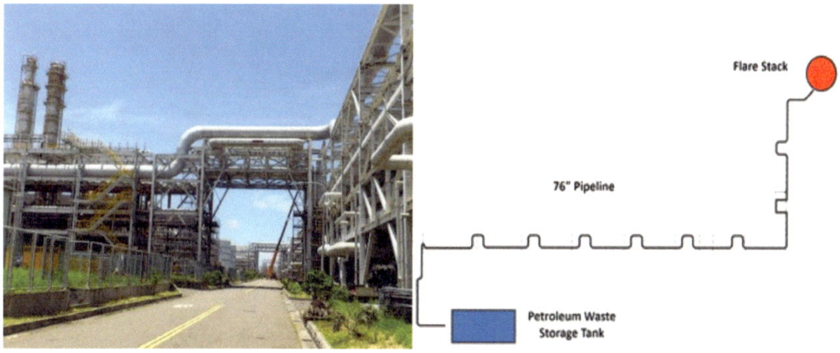

Figure 15.
76-in pipeline with steel frame and plan view [14].

continuous vibrations above the steel frames during the discharge process. The vibration in the steel frames caused by the 76-in pipeline disturbance might lead to cracks in the original pipes below, steel fatigue, and joint failure. Once the above incidents occur, it has potential to result in the escape of poisonous gas, interruption of the production process, and even conflagration. In this case, accelerometer monitoring records are used to detect dynamic structural weaknesses of the steel frames, and then, the structural systems could be retrofitted to reduce the probable essential structure faults leading to industry disasters. Disturbances occur randomly along the 76-in pipeline due to vaporization of solid or liquid waste whose volume expands dramatically and raises the pressure in the pipeline. Waste flow also causes impact force on curved parts of the 76-in pipeline when the flow direction changes.

The IMI 603C01 piezoelectric accelerometer is used in this case. It is a shear-mode-type accelerometer with a ceramic sensing element. It is suggested that ceramic sensing elements provide great resolution and durability in noisy environments and it also covers both low-frequency and high-frequency measurements [15]. Fifty-six accelerometers in either vertical or horizontal direction are installed on the 76-in pipeline and the steel frame below. Accelerometers are aligned vertically along the 76-in pipeline and the steel frame since the response of the steel frame caused by disturbances could be monitored simultaneously by all sensors (**Figure 16**).

With acceleration data from long-term monitoring, locations of vibrations and vibration levels could be identified. To provide methods to reduce vibration, the first step is to build the finite element model verified with monitoring data. In this case, SAP2000 is used to build the finite element model of the steel frame and the 76-in pipeline (**Figure 17**). By assigning time history recorded by the accelerometer, the fundamental frequency of the steel frame could be obtained by the FFT. The fundamental frequency of the steel frame could be also calculated by finite element

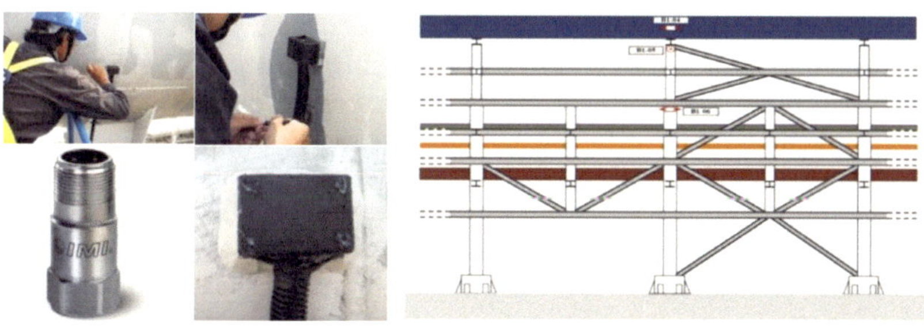

Figure 16.
Accelerometer installation and side view of design [14].

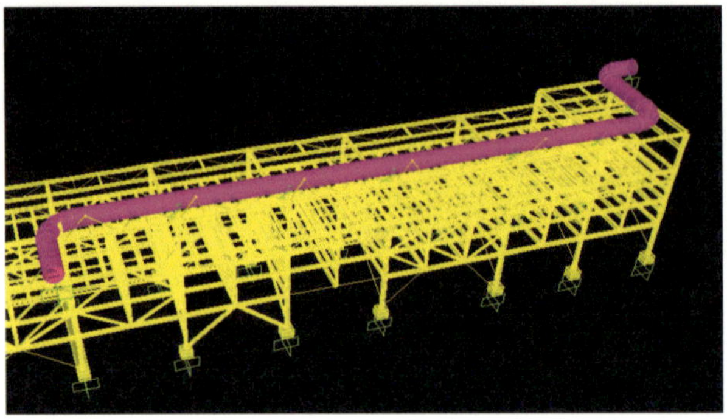

Figure 17.
Finite element model [16].

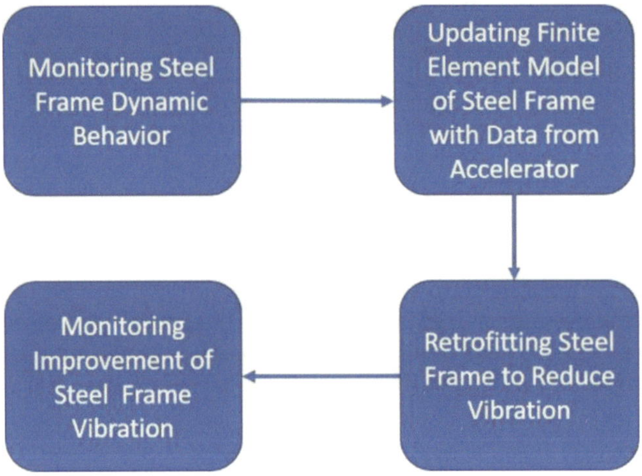

Figure 18.
Research flowchart [16].

software. Therefore, the finite element model could be modified to increase accuracy of the model by comparing frequencies with data recorded by accelerometers. The higher-vibrated steel frame structure is suggested to increase stiffness by installing steel bracing or enlarging column size. Based on the modified finite element model, the effect of retrofitted design could be evaluated in software. After retrofitting, the improvement of the steel frame could be demonstrated by further accelerometers monitoring (**Figure 18**).

The accelerometer plays an important role in this industry case because it provides critical information for steel frame dynamic behavior due to unexpected turbulence. Based on the monitoring data, the accuracy of the finite element model could be enhanced. More accurate models can help structural engineers figure out effective methods to reduce vibration which potentially leads to serious industrial disasters. The improvement could also be validated by further monitoring using accelerometers. On the other hand, steel frame vibration caused by the 76-in pipeline turbulence is also related to the volume of waste delivered to the flare stack. Therefore, the safe range of waste consumption could be determined to avoid insecure vibrations of the steel frame.

9. Conclusion

The purpose of the infrastructural monitoring is to have efficient use of the materials, energy, and labor to increase the performance of infrastructures. Advances of modern remote monitoring increase the efficiency, which is demonstrated in case studies. The emerging sensor technologies, no matter in situ, on-site, or airborne sensors, are increasingly used in the infrastructure sensing. An integrated structural health monitoring system (ISHM) includes the ability to extract information from sensor data to establish trends, such as the sensor signatures and structural damage, and make recommendation of actions to ensure the health of the infrastructures.

Author details

Chung C. Fu*, Yifan Zhu and Kuang-Yuan Hou
The Bridge Engineering Software and Technology (BEST) Center, Department of Civil and Environmental Engineering, University of Maryland, College Park, MD, USA

*Address all correspondence to: ccfu@umd.edu

References

[1] Fu CC, Yunfeng Zhang, Yuan FG. Development of a Self-sustained Wireless Integrated Structural Health Monitoring (ISHM) System for Highway Bridges [Internet]. 2011. Available from: http://ishm.umd.edu/index.php [Accessed: June 10, 2018]

[2] Hsieh KH, Halling MW, Barr PJ. Overview of vibrational structural health monitoring with representative case studies. Journal of Bridge Engineering. 2006;**11**(6):707-715

[3] Farrar CR, Baker WE, Bell TM, Cone KM, Darling TW, Duffey TA, et al. Dynamic Characterization and Damage Detection in the I-40 Bridge Over the Rio Grande. NM (United States): Los Alamos National Lab.; 1994. DOI: 10.2172/10158042. Web

[4] Bendat JS, Piersol AG. Engineering Applications of Correlation and Spectral Analysis. New York: Wiley-Interscience; 1980. 315 p

[5] Fu CC, Wang S. Computational Analysis and Design of Bridge Structures. USA: CRC Press; 2014. ISBN-13 978-1466579842

[6] California Department of Transportation (Caltrans). Available from: http://www.dot.ca.gov [Accessed: June 10, 2018]

[7] Michael Minor & Associates (Oregon, USA). Available from: http://www.drnoise.com [Accessed: November 9, 2018]

[8] Andrews J, Buehler D, Gill H, Wesley L. Transportation and Construction Vibration–Guidance Manual, (CT-HWANP-RT-13-069.25.3). CALTRAN; 2013

[9] Wikipedia. Acceleration [Internet]. Available from: https://en.wikipedia.org/wiki/Acceleration#/media/File:1-D_kinematics.svg [Accessed: June 10, 2018]

[10] Fu CC, Alayed H, Amde AM. Field performance of the fiber reinforced polymer (FRP) deck of a truss bridge. United States: Journal of Performance of Constructed Facilities, USA: ASCE. 2007;**21**(1 Jan/Feb):53-60

[11] Fu CC. Load Test Report—Bridge No. 12016 on MD-24 over Deer Creek (FRP Bridge Deck Replacement) to Maryland State Highway Administration and Federal Highway Administration IBRC project. Rev. 2. 2003. Available from: https://cpb-us-e1.wpmucdn.com/blog.umd.edu/dist/f/392/files/2016/08/MD-24-Report_R2-2egkdtr.pdf [Accessed: June 10, 2018]

[12] Li X, Sun M, Fu CC. Fast assessment of hanger cables on arch-girder bridges. In: Bridge Maintenance, Safety, Management and Life Extension Chapter 210. London: CRC Press/Taylor & Francis Group. 2014. pp. 1527-1534

[13] Mahoney D. Large property damage losses in the hydrocarbon-chemical industries: A thirty-year review. M & M Protection Consultants; 1997

[14] Yi-Chen Z, Tsung-Chin H. A field study of dynamic measurement and security assessment of large-scale petrochemical structures [thesis]. Taiwan: Civil Engineering, National Kaohsiung University of Science and Technology; 2013

[15] PCB. Model 603C01 Platinum Low-cost Industrial ICP® Accelerometer Installation and Operating Manual; 2010

[16] Tsung-Hsuan L, Tsung-Chin H. Dynamic simulation and strengthening analysis of large steel frame structure [thesis]. Taiwan: Civil Engineering, National Cheng Kung University; 2014

Chapter 7

Geo Spatial Analysis for Tsunami Risk Mapping

Abu Bakar Sambah and Fusanori Miura

Abstract

Tsunami risk is a combination of the danger posed by tsunami hazard, the vulnerability of people to an event, and the probability of destructive tsunami. The spatial multicriteria approach made a possibility for integrating the vulnerability and risk parameters to assess the potential area that will be affected by the tsunami. The study applied the parameters of physical and social vulnerability and combined element at risk to assess tsunami risk in the coastal area of East Java Indonesia. All parameters in both tsunami vulnerability and tsunami risk assessment were analyzed through cell-based analysis in geographical information system. The weight of each parameter was calculated through the analytical hierarchy process. The results were provided as maps of tsunami vulnerability and tsunami risk. Tsunami risk map described five classes of risk. It described that coastal area with a low elevation and almost flat identified as high risk to the tsunami. The coastal area with a high density of vegetation (mangrove) was defined as the area with low level of tsunami risk. The existence of river and other water canals in coastal area was also analyzed for generating tsunami risk map. Risk map highlights the coastal areas with a strong need for tsunami mitigation plan.

Keywords: tsunami, vulnerability, risk, geospatial, weighted overlay, GIS

1. Introduction

Tsunami can be defined as a series of waves created by an impulsive disturbance in the water body. It causes severe damage to coastal areas. A tsunami wave could be less than 1 m high in the open ocean and traveling at up to 800 km/h in which the wave energy will be extended from the surface to the ocean floor. The wave energy of tsunami will be compressed into a much shorter distance when it approaches the coast, creating potentially large destructive to the coastal areas [1]. A tsunami can be generated when the sea floor abruptly deforms and a bottom layer of water body displaces the overlying water vertically. One kind of earthquake that is related to the crustal deformation of the earth is tectonic earthquakes. When these earthquakes happen in the bottom of the sea, the water layer above the deformed area is displaced from its equilibrium position. Waves are formed as the displaced water mass, which occurs due to the impact of gravity. A tsunami can be generated when large areas of the sea floor subside.

In the deep water of the open ocean, the speed of tsunami waves can be up to 800 km/h. The energy wave of tsunami will decrease dramatically when it approaches the coast, but its height can be 10 times or more and have catastrophic consequences to the coastal areas. As a result, the low-lying areas of the coast and the areas near bay mouths or tidal flats will be highly vulnerable to the tsunamis. A tsunami can be generated through four special events, which are illustrated in **Figure 1**. These figures described that the distribution of tsunami was based on the effect and magnitude, which describes that mostly this event occurred in the area of "ring of fire," a zone of active earthquakes and volcanoes, surrounds much of the Pacific Ocean.

1. Subduction zone

One of the many plates that make up earth's outer shell descends under an adjacent plate. This kind of boundary is called a subduction zone. When the plates move suddenly in an area where they usually stick, an earthquake will happen.

2. Between earthquakes

Stuck to the subducting plate, the overriding plate gets squeezed. Its leading edge is dragged down, while an area behind bulges upward. This movement goes on for decades or centuries, slowly building up stress.

3. During earthquakes

An earthquake along a subduction zone happens when the leading edge of the overriding plate breaks free and springs seaward, raising the seafloor and water above it. This uplift starts a tsunami. Meanwhile, the bulge behind the leading lead collapses, flexing the plate downward and lowering the coastal area.

4. After earthquakes

Part of the tsunami races toward nearby land, growing taller as it comes into shore. Another part heads across the ocean toward distant shores.

The 2004 Indian Ocean earthquake and tsunami, the 2011 Tohoku earthquake and tsunami, and the 2018 Sulawesi Indonesia earthquake and tsunami show that

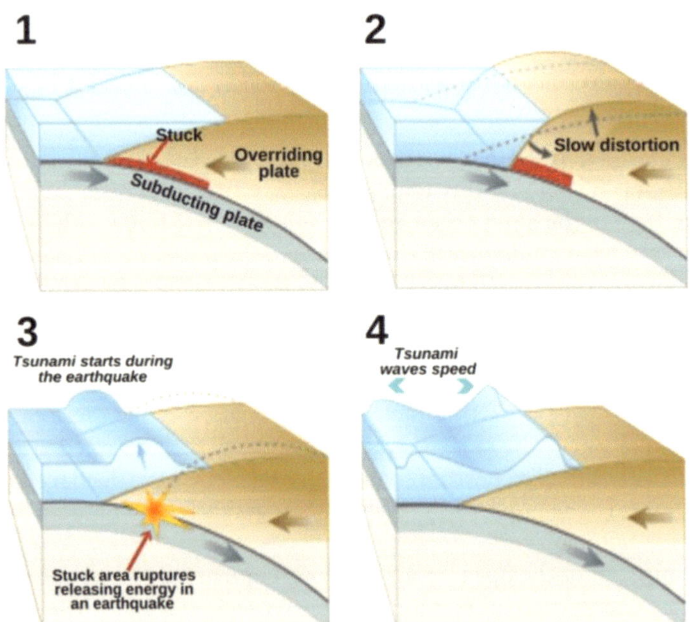

Figure 1.
Tsunami generations [2].

this natural disaster is almost periodical event in 7–10 years along the area of ring of fire, a major area in the basin of the Pacific Ocean where many earthquakes and volcanic eruptions occur. The natural disaster includes tsunami is the even that it is impossible to reduce its occurrence, but the impact can be minimized by performing an initial assessment related to vulnerability and risk mapping.

Tsunami risk mapping combines the results of the tsunami vulnerability and tsunami hazard. This illustration is described in **Figure 2**. Assessing tsunami vulnerability and risk can provide important information for tsunami mitigation plan. This also plays an important role in preparing and mitigating for the future events of tsunami [3, 4]. A risk of a tsunami disaster is defined as the mathematical product of tsunami vulnerability and tsunami hazard. It refers to the expected loss from a given hazard to a given element at risk [5]. A disaster is a function of the risk process. Risk results from the combination of hazards, vulnerability, and insufficient capacity to minimize the negative impact of risk. In general, risk assessment combines the results of the hazard and vulnerability assessments [6].

Moreover, disaster risk assessment is a qualitative or quantitative approach to determine the nature and extent of disaster risk by analyzing potential hazards and evaluating existing conditions of exposure and vulnerability that together could harm people, property, services, livelihoods, and the environment on which they depend [7].

The implementation of tsunami risk assessment includes the effectiveness of mitigation will be the practical outcome of the study. Risk map can be used for both tsunami evacuation route and tsunami evacuation building setting. In order to create tsunami risk map, it is necessary for assessing the vulnerability areas due to tsunami in which the physical and social parameters of vulnerability are needed.

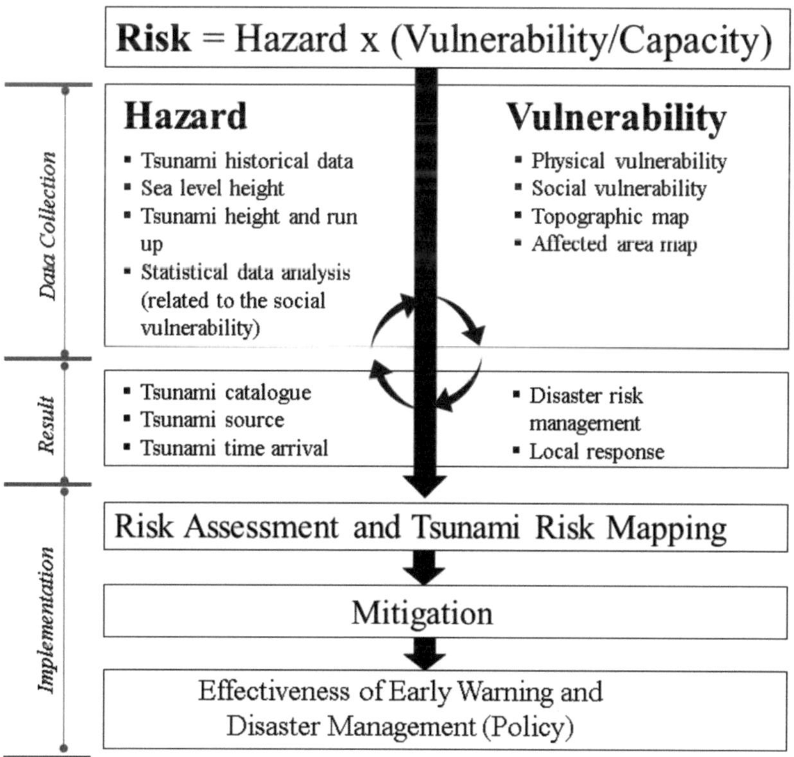

Figure 2.
General concept of the study.

Many approaches have been applied in order to map the potential areas affected by tsunami. A novel approach of tsunami vulnerability mapping in the application of geographical information system (GIS) together with the analysis of satellite remote sensing has been done to assess the risk areas due to tsunami. An overview of the use of geospatial data with emphasis on satellite remote sensing data and of the approaches used for hazard assessment is given. Satellite images have the advantage of being able to deliver simultaneous images of wide areas [8–11]. Remote sensing imagery is already used for hazard-related applications since the advent of research-oriented satellite systems and sensors four decades ago.

In addition, with the aid of the GIS, geospatial analysis helps prioritize the decision-making process using georeference data. Geospatial analysis through the application of spatial multicriteria analysis is vastly different from conventional multicriteria decision-making techniques, due to the inclusion of an explicit geographic element. Spatial multicriteria analysis uses information on both the criterion values and the geographical information, in addition to the decision-maker's preferences with respect to a set of evaluation parameters [12, 13].

Some of the previous studies on tsunami vulnerability have analyzed remote sensing data, primarily to assess the physical vulnerability and risk of coastal areas. In addition to such studies, the application of remote sensing in hazard and vulnerability assessment related to ecological and socioeconomic vulnerability has been analyzed. Previous studies have also applied moderate-resolution optical satellite images and integrated analysis using GIS to identify inundation areas due to tsunamis [14–17]. GIS mapping of tsunami vulnerability has also applied using the Shuttle Radar Topography Mission (SRTM) to obtain the topographic data of the study area [18]. Another spatial analysis method has applied soil type, urban form, and social type system for the potential natural hazard mapping [Hsien] and has determined the tsunami-vulnerable area by comparing building damage map with the topography data, which is discussed with regard to land elevation, land use, and the distance from the coast [19].

Mapping of the 2011 Tohoku Earthquake tsunami inundation and run-up by survey also has been published [20]. A novel approach from the Coastal Risk Analysis for Tsunamis and Environmental Remediation (CRATER) project was applied for assessing tsunami vulnerability on a regional scale using ASTER imagery and SRTM version 3. This work analyzed the vulnerability of coastal zones and inland areas using the parameters of infrastructural, geomorphological, and ecological features for coastal zones, and parameters of land use, altimetry, and distance from the shoreline for inland areas [21]. Tsunami vulnerability mapping along coastal area of East Java also applied using high resolution of elevation data from NEXTMap World 10.

The main contribution of this book chapter is to utilize the remote sensing technology based on geospatial technique to investigate the spatial tsunami risk impact on coastal zones. In this regard, the main objective is to establish a geospatial tool for monitoring the tsunami risk impacts along the coastal area of East Java, Indonesia.

2. Method

2.1 Study area

The study was applied at the south coastal area of East Java, Indonesia (**Figure** 3). The coastal areas of East Java including Malang district and Jember district were known as one of the important marine fishery resource spots in East Java. These areas

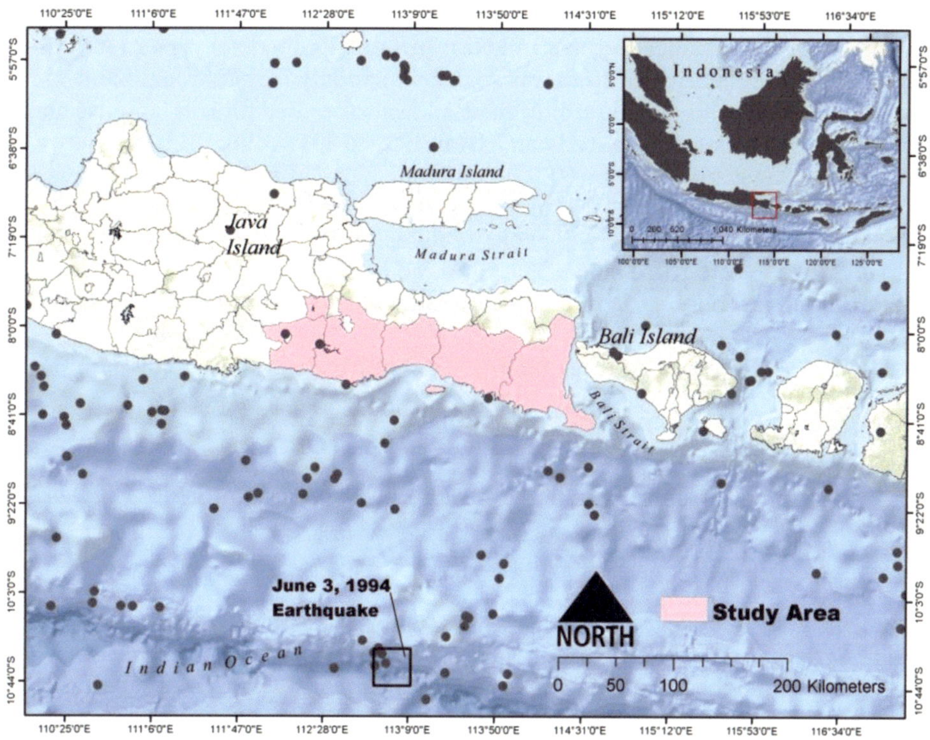

Figure 3.
Study area was the coastal area of East Java. [Points represent historical data of earth quake epicenter].

also affected by 1994 tsunami event along the coastal area of East Java. The tsunami genic earthquake occurred on June 3, 1994, in the Indian Ocean about 200 km south of Java. The earthquake, which had a surface-wave magnitude of 7.2 and a moment magnitude of 7.8 at 10.51°S and 112.87°E, generated a devastating tsunami that took the lives of more than 200 East Java coastal residents; with maximum run-up value of 9.50 m measured at Rajekwesi area, east part of the study area [22, 23].

2.2 Dataset

Tsunami vulnerability map was created using two deference sources of elevation data. Digital Elevation Model (DEM) from the ASTER Global Digital Elevation Model (ASTER GDEM) version 2 was applied to generate elevation and slope map. The Advanced Spaceborne Thermal Emission and Reflection Radiometer (ASTER) GDEM is a joint product developed and made available to the public by the Ministry of Economy, Trade, and Industry (METI) of Japan and the United States National Aeronautics and Space Administration (NASA).

DEM of NEXTMap World 10 also applied for creating the elevation and slope parameter as a digital elevation data for physical tsunami vulnerability mapping. NEXTMap World 10 provides a 10 m resolution and makes the wide possibility for elevation analysis in large areas. This DEM was the combination process from SRTM90 v2.1: 90 m Digital Surface Model (DSM) collected from Interferometric Synthetic Aperture Radar (IFSAR) in February 2000, SRTM30: 30 m DSM also collected from IFSAR in February 2000, ASTER 30v2.0: 30 m DSM collected from optical sensor from 2005 to 2011, ICESat as a point data of LiDAR from Geoscience Laser Altimeter System (GLAS) satellite collected from 2003 to 2010, and GTOPO30: 30 m DSM collected from eight raster and vector data from USGS in 1996.

In order to map the land use of the study area, ALOS satellite imagery with the instrument of the Advanced Visible and Near Infrared Radiometer type 2 (AVNIR-2) with the spatial resolution of 10 m was analyzed. Landsat 8 Operational Land Imager (OLI) with the spatial resolution of 30 m also applied for this land use map. Moreover, seismic data of the study area from 1992 to 2014 collected from the United States Geological Survey (USGS), and downloaded from http://earthquake.usgs.gov/earthquakes/search/ was used as a supporting parameter for tsunami vulnerability in which further will generate the seismic map. In order to apply the overlay process, vector base map of East Java Indonesia was used to prepare vector data of coastal morphology, coastal line (coastal proximity), and river proximity.

2.3 Satellite image processing for land cover classification

2.3.1 Digital number to radiance conversion

The algorithm that applied for the conversion of DN to radiance applied Eq. (1) [24].

$$L_\lambda = G_{rescale} \times QCAL + B_{rescale} \tag{1}$$

in which L_λ is the spectral radiance at the sensor's aperture (W/m^2/sr/μm), $G_{rescale}$ is the rescaled gain, $QCAL$ is the digital number (DN), and $B_{rescale}$ is the rescaled bias. **Table 1** described rescaled gains and biases for ALOS AAVNIR-2 satellite.

2.3.2 Radiance to reflectance conversion

The conversion of radiance to reflectance applied the algorithm of Eq. (2) [25].

$$\rho_\lambda = \pi \times L_\lambda \times d^2/ESUN_\lambda \times \cos\theta_s \tag{2}$$

where ρ_λ is the unitless planetary reflectance, L_λ is the spectral radiance at the sensor's aperture, d^2 is the earth-sun distance in astronomical units from a nautical handbook, $ESUN_\lambda$ is the mean solar exoatmospheric irradiance, and θ_s is the solar zenith angle in degree.

2.3.3 Decision tree classification

Satellite image classification was applied a decision tree classification. The value that was used for the range of classification was based on the normalized difference vegetation index (NDVI) value. NDVI is a measure of the difference in reflectance between these wavelength ranges with the values from −1 to 1. NDVI value more than 0.5 indicates dense vegetation and the value less than 0 indicates no vegetation including water. NDVI was calculated using Eq. (3) [26].

Band	$G_{rescale}$	$B_{rescale}$
1	0.5888	0
2	0.5730	0
3	0.5020	0
4	0.8350	0

Table 1.
Rescaling gains and biases used for DN to spectral radiance conversion (for ALOS AVNIR-2).

$$NDVI = \frac{(NIR - VIS)}{(NIR + VIS)} \quad (3)$$

NIR is the reflectance of near-infrared band and *VIS* is that of visible red band of satellite sensor. Band 3 is represented in red and band 4 is *NIR*.

Decision tree classification is a flowchart like a tree structure where each internal node donates a test on an attribute and each branch represents an outcome of the test and leaf nodes represent the class distribution [27]. The concept of decision tree classification is illustrated in **Figure** 4. The NDVI value was applied as a basic expression. Each decision is based on a numerical comparison with a selected threshold index, which makes the whole process easily repeatable. The decision tree is also applied in the study of tsunami vulnerability using ASTER imagery [21]. The main advantage of such approach is that data from many different sources can be processed together to make a single decision tree classifier. Decision tree tool is nonparametric; therefore, it makes no assumption on the distribution of the input data [28]. The NDVI represents a simple numerical indicator that can be used in analyzing remote sensing imagery and assessing whether the target being observed contains live green vegetation.

Decision tree classification step applied as below.

1. Entering the rules

    a. The decision tree tool starts with one empty decision node that will divide the pixels in the dataset into two groups using binary decision expression entered into that empty node.

    b. The first decision will be based on the medium resolution of satellite image. The decision can be created by adding the decision node labeled.

    c. Expression of the decision can be set in the decision node based on the criteria of analysis (for example, NIR < 0.09). The text will appear in the decision node in the graphical view of decision tree.

    d. The variable pairings dialog appears after adding the expression.

2. Pairing the expression variable with a file

In this step, an expression that already set will be linked to the associate file. This describes the decision tree that when evaluating this decision expression, the expression should be calculated from the associate file.

2.4 Data processing and analysis by modeling tsunami physical, social, and risk vulnerability

2.4.1 Physical and social vulnerability

Vulnerability mapping has been generated using the parameter of elevation, slope, coastal proximity, river proximity, coastal morphology (illustrated coastal type), and land use. Together with hazard or capacity, tsunami vulnerability is one of the parameters in assessing tsunami risk. The cell-based analysis was applied in combining all parameters through GIS process. Each tsunami vulnerability parameter was classified into five classes for vulnerability based on the criteria range as shown in **Tables 1** and **2**. The coastal proximity classes in meter were calculated

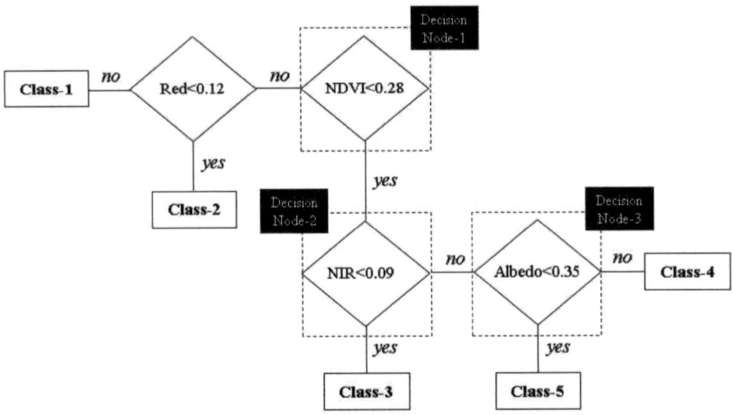

Indices used in decision tree	
NDVI	NDVI < 0.28
Red	RED < 0.12
NIR	NIR < 0.09
Albedo	Albedo < 0.35

Figure 4.
Illustration of decision tree applied in satellite image classification.

based on the measured run-up and water height in the surveyed area during the last tsunami event in the coastal area of East Java on June 3, 1994. It was calculated using algorithm Eq. (4) [29, 30].

$$\log X_{max} = \log 1400 + 4/3 \log (Y_0/10) \quad (4)$$

X_{max} is the maximum reach of the tsunami over land, and Y_0 is the height of the tsunami at the coast.

Moreover, land use map was generated from supervised classification process of satellite mage. Maximum likelihood methods were applied in the supervised process. Land use class was divided into five classes, and each class was reclassified based on the vulnerability classes as shown in **Tables 2** and **3**. The selected sample for each land use class in the reflectance value of satellite digital image was collected to calculate the accuracy of classification result. The classification of land use was based on the spectral signature defined in the training set.

Vulnerability class	Physical vulnerability		
	Elevation (m)[1]	Slope (%)[2]	Land use[3]
High	<5	0–2	Urban
Slightly high	5–10	2–6	Agriculture
Moderate	10–15	6–13	Bare soil
Slightly low	15–20	13–20	Water
Low	>20	>20	Forest

[1]*Ref. [31].*
[2]*Ref. [32].*
[3]*Ref. [30].*

Table 2.
Physical tsunami vulnerability value range (1).

Vulnerability class	Physical vulnerability		
	Coastal distance (m)[1]	River proximity (m)[2]	Coastal type[2]
High	<293	0–100	V bay
Slightly high	293–514	100–200	U bay
Moderate	514–762	200–300	Cape
Slightly low	762–1032	300–500	Straight
Low	>1032	>500	Neutral

[1]*Based on Eq. (4) calculations.*
[2]*Ref. [33].*

Table 3.
Physical tsunami vulnerability value range (2).

In the term of social vulnerability analysis, a social parameter was needed. Social vulnerability can be defined as the exposure of groups or individuals to unexpected changes and disruption to livelihoods [34]. Social vulnerability also can be measured as a result of social and place inequalities [35]. Social vulnerability is defined also as the limitation of a community to the impact of natural disasters that influence its ability or resilience in order to mitigate and recover from and prepare for the impacts of disaster [36]. Social vulnerability map was created using four parameters and weighted equally based on the criteria as explained in **Table 4**.

2.4.2 Spatial multicriteria analysis

The parameters of both physical and social vulnerabilities are displayed in grid cells, which are then classified based on their value to five classes of vulnerabilities; they represent low, slightly low, medium, slightly high, and high vulnerability. All parameters will be overlaid in the raster data format (cell-based) based on their weight. Weighted overlay describes the technique for applying a common measurement scale of values to diverse and dissimilar inputs to create an integrated analysis. Weighted overlay also describes the type of suitability analysis that helps in analyzing site conditions based on multiple criteria. By identifying areas based on their criteria, weighted overlay analysis allows the user to combine weight and rank several different types of information and give the visualization of the result, in which multiple parameters can be evaluated at once [38]. Weights for all parameters are constructed in terms of pair-wise comparison matrix through analytical hierarchy process (AHP).

AHP can be defined as an approach for organizing and analyzing complex decisions, based on mathematics (matrix calculation) and psychology. In this study, AHP helps in constructing the weight of each parameter by applying expert judgment. The result from AHP calculation then overlaid spatially in GIS methods. The concept of pair-wise comparison and AHP calculation is illustrated in four steps below.

Step 1. Construct pair-wise comparison matrix, in which each parameter will be compared to others as described below.

	c1	c2	c3	c4	c5
c1	1	c_1/c_2	c_1/c_3	c_1/c_4	c_1/c_5
c2	c_2/c_1	1	c_2/c_3	c_2/c_4	c_2/c_5
c3	c_3/c_1	c_3/c_2	1	c_3/c_4	c_3/c_5

c4	c_4/c_1	c_4/c_2	c_4/c_3	1	c_4/c_5
c5	c_5/c_1	c_5/c_2	c_5/c_3	c_5/c_4	1
sum	S_1	S_2	S_3	S_4	S_5

c1: parameter 1 c4: parameter 4
c2: parameter 2 c2: parameter 5
c3: parameter 3

Step 2. Normalized matrix

$$\begin{bmatrix} (c1/c1)/s1 & (c1/c2)/s1 & (c1/c3)/s3 & .. & (c1/c5)/s5 \\ (c2/c1)/s1 & (c2/c2)/s1 & (c2/c3)/s3 & .. & (c2/c5)/s5 \\ (c3/c1)/s1 & (c3/c2)/s1 & (c3/c3)/s3 & .. & (c3/c5)/s5 \\ \cdots & \cdots & \cdots & .. & \cdots \\ (c5/c1)/s1 & (c5/c2)/s1 & (c5/c3)/s3 & .. & (c5/c5)/s5 \end{bmatrix} \begin{matrix} \rightarrow \\ \rightarrow \\ \rightarrow \\ \vdots \\ \rightarrow \end{matrix} \begin{bmatrix} s1.1 \\ s1.2 \\ s1.3 \\ \cdots \\ s1.5 \end{bmatrix} \begin{bmatrix} n1 \\ n2 \\ n3 \\ \cdots \\ n5 \end{bmatrix}$$

n5: fifth iteration of principal eigenvalue

Step 3. Calculation of consistency ratio

$$CR = \frac{CI}{RI} \tag{5}$$

$$CI = \frac{(\lambda_{max} - N)}{(N - 1)} \tag{6}$$

in which, CR is consistency ratio, CI is consistency index, RI is random consistency index, λ_{max} is the principal eigenvalue, N is the number of the comparison matrix.

RI values depend on matrix size (N) as explained below.

Matrix size (N)	1	2	3	4	5	6	7	8	9	10
RI	0	0	0.58	0.90	1.12	1.24	1.32	1.41	1.45	1.49

If the value of CR in less than or equal to 0.10 (10%), the inconsistency is acceptable. If the CR is more than 10%, then the subjective judgment needs to be revised.

In AHP processing, pair-wise matrix was done to compare the importance of each parameter using Saaty's nine scales. In the second step of AHP, the normalized matrix was applied to calculate the eigenvalue of the parameter in which the parameter's weight was constructed. Eigenvalue represents the weight of each parameter, and it is applied after the fifth iteration of principal eigenvalue. Consistency ratio (CR)

Parameters (a)	Σ (b)	Proportion* (c)	Score** (d)
Population density	P	(b)/total population	(c)/maximum proportion
Gender	G	(b)/total woman	(c)/maximum proportion
Age***	A	(b)/total age	(c)/maximum proportion
Disabilities	D	(b)/total disabilities	(c)/maximum proportion

**Determine the factor of each village divided by number per subdistrict.*

***The same value for all places on all the social variables.*

****Number of elderly and children.*

Table 4.
Social vulnerability parameter [37].

calculation was done to determine the inconsistency from AHP process in which RC should be less than 10%. Weighted linear combination is very straightforward in a raster GIS, and factors are combined by applying a weight value to each followed by a summation of the results to create both physical and social vulnerabilities using Eq. (7) [36] and Eq. (8).

$$\text{Vulnerability} = \Sigma(W_i \cdot X_i) \tag{7}$$

$$\text{Total vulnerability} = \Sigma(\text{physical vulnerability} \times \text{weight}) + (\text{social vulnerability} \times \text{weight}) \tag{8}$$

where W_i is the weight value of the parameter i and X_i is the potential rating of the factor.

The illustration of weighted/cell-based overlay is described in **Figure 5**.

2.4.3 Risk assessment

Risk combines the vulnerability result with the probable level of loss to be expected from a predictable magnitude of hazard (which can be considered as the manifestation of the means that produces the loss). Risk, vulnerability, and hazard are the three factors or elements, which we are considering here in this pseudoequation. The terminology of risk also given by factor analysis of information risk which may be related to disaster is "the probable frequency and probable magnitude of

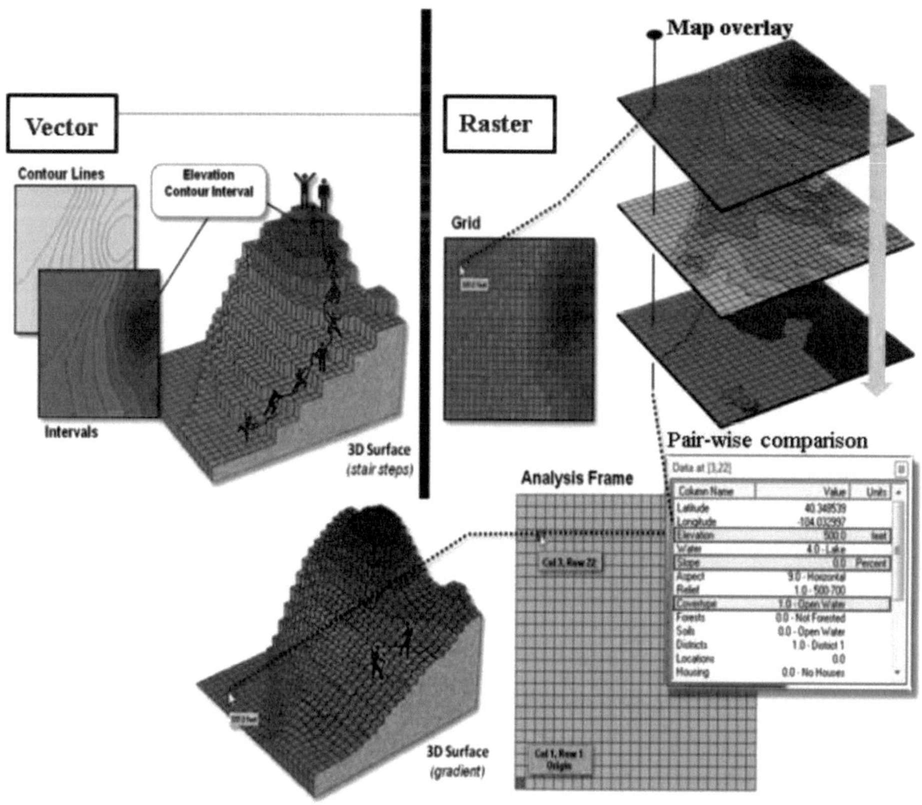

Figure 5.
Weighted/cell-based overlay illustration.

predicted losses" [34]. The value of risk can be calculated as the product of vulnerability and hazard level. Vulnerability level ranges from 1 to 5 from low to high vulnerability and hazard level ranges from 1 to 4. Risk was calculated using Eq. (9).

$$R = V \times {}^{H}\!/_{4} \tag{9}$$

in which R is a risk, V is vulnerability, and H is a hazard.

R must be an integer number ranging from 1 to 5, where 5 stands for the maximum risk level. Once risk level has been calculated, it will be possible to plot it on a risk map by the process of GIS. Risk map was also generated using weighted cell-based overlay. Weighted overlay analysis allows the user to combine weight and rank several different types of information and visualize it, so multiple factors can be evaluated at once [38].

3. Result and discussion

3.1 Image classification for land use mapping

Land use is one of the parameters that is applied in the spatial multicriteria analysis in creating physical tsunami vulnerability map. Land use map was created using decision tree classification in which the concept was as illustrated in **Figure 4**. The result of the image classification is described in **Figure 6**.

In order to create a tsunami vulnerability map based on the land use, a category of vulnerability classes was applied in the land use raster map. Tsunami vulnerability map was created using the criteria as described in **Table 1**, in which it briefly described

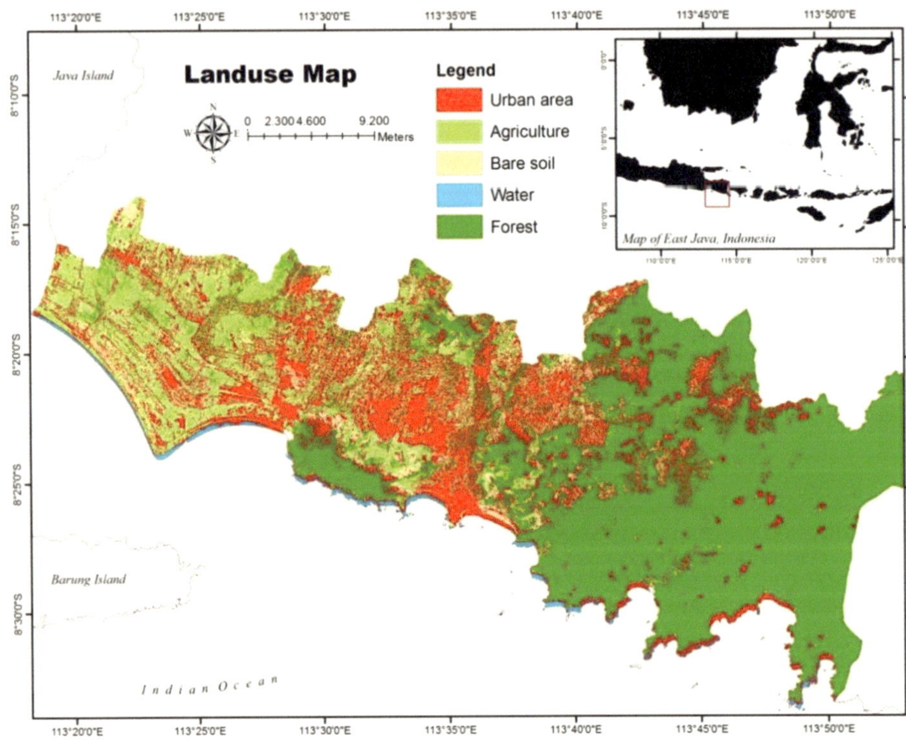

Figure 6.
Land use of the study area as the result of satellite image processing.

that high class of vulnerability was set for the land use of urban area, slightly high for the agriculture, moderate for the bare soil area, slightly low for the water body, and low vulnerability class for the forest area. The reclassification of land use based on those criteria created the tsunami vulnerability map as shown in **Figure 7**.

Mostly, urban area was spread in the western part along the coastal of study area. These areas are known as a center of fishery activities with many fishing bases. An urban area spread in the low elevation until flat area, and it will get serious impact when tsunami wave approaches the coastal area. In the area with high density of coastal vegetation, the damage impact of the tsunami will be lower. Coastal vegetation, such as mangrove, can act as the barrier zone to reduce the energy of tsunami wave when it hits the coastal area, and it will also minimize the impact of tsunami.

3.2 Tsunami vulnerability mapping

Physical vulnerability map was created using six criteria with different weight. The weight of each criterion was calculated using pair-wise comparison matrix in the AHP approach. Some experts in the theme of tsunami prevention system and management were collected to construct the matrix of criteria. Pair-wise comparison matrix and the result calculation of CR for the physical vulnerability are described in **Table 5**.

Moreover, social vulnerability was calculated based on social database in the theme of population density, gender, age, and also education level. The map of these criteria is based on scoring and weight of each criterion, which will be basic information in determining both evacuation route and evacuation building. The calculation was based on [39] in which the first step in calculation was calculation of X for determining the percentage of woman. This calculation was using Eq. (10).

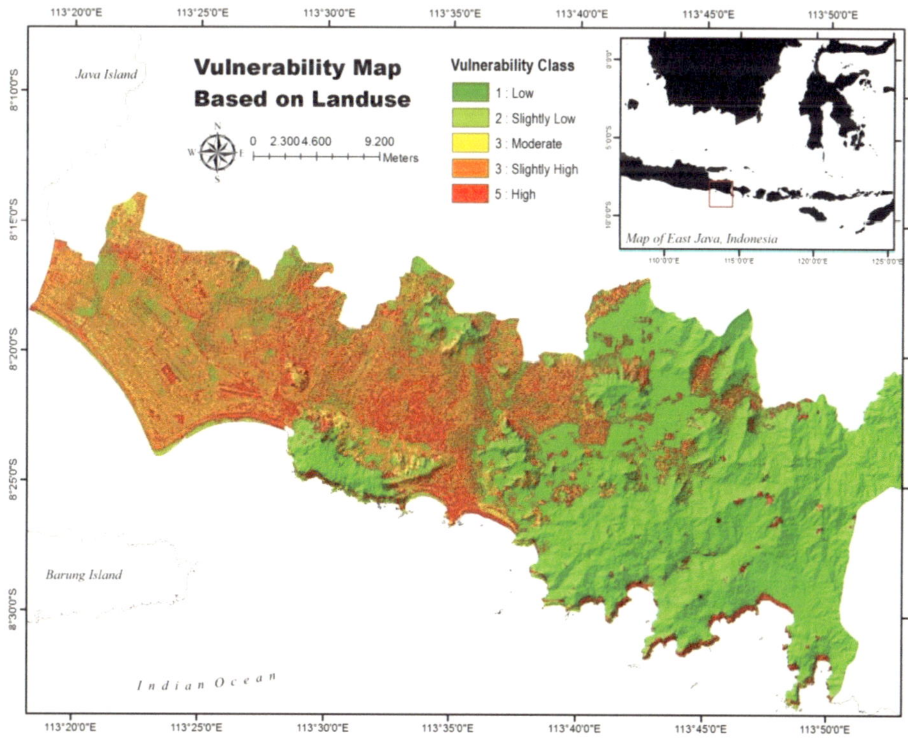

Figure 7.
Tsunami vulnerability map based on land use.

The second step was to calculate the score of woman by calculating the ratio of X for determining scale using Eq. (11)

$$X = \frac{\text{total of woman in district}}{\text{total of woman in study area}} \tag{10}$$

$$\text{score for woman} = \frac{X}{\text{maximum } X} \tag{11}$$

Scoring of the parameter was calculated using **Table 6**.

By using Eq. (8), tsunami vulnerability map was created. The map of tsunami vulnerability is shown in **Figure 8**, which illustrated that from the analysis of both physical and social parameters of vulnerability, the high class of tsunami vulnerability area is found in the area with low elevation along the coastal area. The coastal area vulnerable to tsunami inundation, the urban area, and infrastructure are not uniformly at risk within the flood zone [35]. The affected area due to tsunami is related both to vulnerability and to the tsunami wave energy; it is also related to the source of earthquake epicenter. Damage level to buildings in the urban area depends on building type and the height of the building and on inundation depth [37] or could depend on vegetation density around the coastal area. Mangrove can be a buffer zone for the impact of tsunami wave and it is assumed that it will reduce tsunami impact.

3.3 Seismic analysis and tsunami run-up

Seismic data consist of physical measurements, seismic sources, seismic waves, and their propagating media. The purpose of seismic data processing is to learn something about the earth's interior. It needs to figure out some specific relations between the intended targets and measurable parameters in order to understand certain aspects of the earth [38]. All initial tsunami warnings are based on early detection and characterization of seismic activity. Due to the fundamental differences in nature between the solid earth in which an earthquake takes place and the fluid of ocean where tsunami gravity waves propagate, the vast majority of earthquakes occurring on a daily basis do not trigger appreciable or even measurable tsunamis. It takes a large event (magnitude more than 7.0) to generate a damaging tsunami in the near field and a great earthquake (magnitude more than 8.0) to generate a tsunami in the far field [39].

	Normalized principal eigenvector (fifth iteration)						
	c.1	**c.2**	**c.3**	**c.4**	**c.5**	**c.6**	**%**
c.1	0.28	0.29	0.353	0.288	0.184	0.273	28
c.2	0.187	0.194	0.176	0.231	0.184	0.182	19
c.3	0.14	0.194	0.176	0.231	0.245	0.136	18
c.4	0.112	0.097	0.088	0.115	0.184	0.182	12
c.5	0.187	0.129	0.088	0.077	0.122	0.136	13
c.6	0.093	0.097	0.118	0.058	0.082	0.091	9

CI = 0.032, CR = 2.6%.
c.1: elevation; c.2: slope; c.3: coastal distance; c.4: river proximity; c.5: coastal type; c.6: land use.

Table 5.
Pair-wise comparison matrix.

Parameter (a)	Proportion (b)	Score (c)	Weight (d)
Population density	(b)/total population	(c)/maximum proportion	25
Gender	(b)/total woman	(c)/maximum proportion	25
Age	(b)/total age	(c)/maximum proportion	25
Disabilities	(b)/total disabilities	(c)/maximum proportion	25

Table 6.
Weight of the parameter for social vulnerability [39].

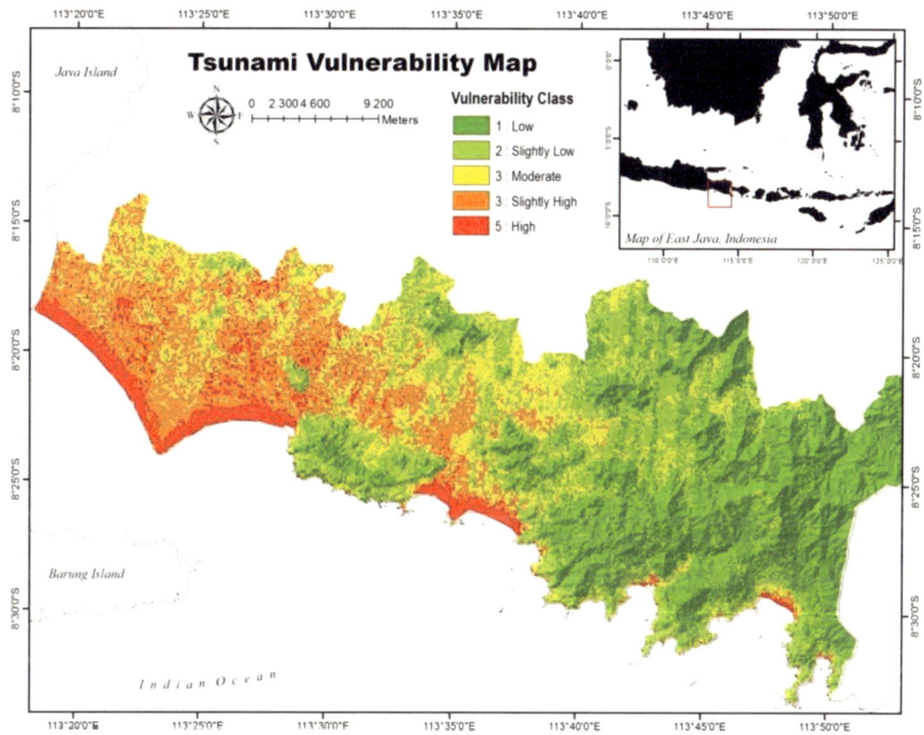

Figure 8.
Tsunami vulnerability map.

Coastal area of East Java was identified as the area with high number of the seismic point with the depth of 1–5 km in average and the range of magnitude was 1 to more than 6. As the result, the coastal area of East Java can be classified as a high risk to tsunami. Based on historical event, it was a magnitude of 7.8 Mw in the depth of 18 km that identified the geographic position in latitude of −10.477° and longitude of 112.835°, caused a big tsunami, and affected the coastal area of East Java.

Tsunami run-up parameter plays as an important role in determining tsunami risk. This parameter is also classified as a main parameter in hazard criteria. The historical tsunami event close the coastal area of East Java in 1994 was used as basic data for run-up analysis. The maximum run-up was recorded around 11.2 m at Tempurejo district, and minimum run-up was 3.1 m in the area of Puger district. The run-up analysis described seven run-up points along the coastal area. The last survey described that in the North West part of the study area (Cape Pelindu), a small fishermen village where a fishery created a sort of barrier to the sea water, separating houses from the ocean [37]. The fishery defense wall and three typical straw houses were destroyed. According to eyewitnesses, three big waves followed each other, the third one being the biggest. The measured maximum water height was 3.20 m, and the maximum water ingression was about 350 m.

3.4 Tsunami risk mapping

Tsunami risk map describes five classes of risk level. Moderate until the high class of tsunami risk was found in the western part of the study area. This area is mostly covered by flat elevation, bare soil, and the area that contains rare density of coastal vegetation and high density of building area. Previous tsunami event recorded that run-up of 4.85–5.85 m happened in this area.

Tsunami risk map was created by the spatial integration of vulnerability and hazard map. Tsunami risk can be defined as a combination of the danger posted by tsunami event, the vulnerability of people to tsunami hazard, and the probability of destructive *tsunami*. Tsunami risk as the result of this study is described in **Figure 9**. Tsunami waves may undergo extensive refraction and create a process that may converge their energy to particular areas on the coastal area and increase the heights and run-up of the waves when they hit the coastal area. High risk of a tsunami is depending on the depth of water, the coastal geomorphology, the direction of the tsunami wave, and the existence of rivers or other water canals.

The high density of mangrove in coastal areas and the existence of reefs can play as a barrier to reduce the effect of the tsunami wave, as well as the islands with steep-sided fringing are only at moderate risk from tsunamis. Study about predicting tsunami inundation area using coastal vegetation density was carried

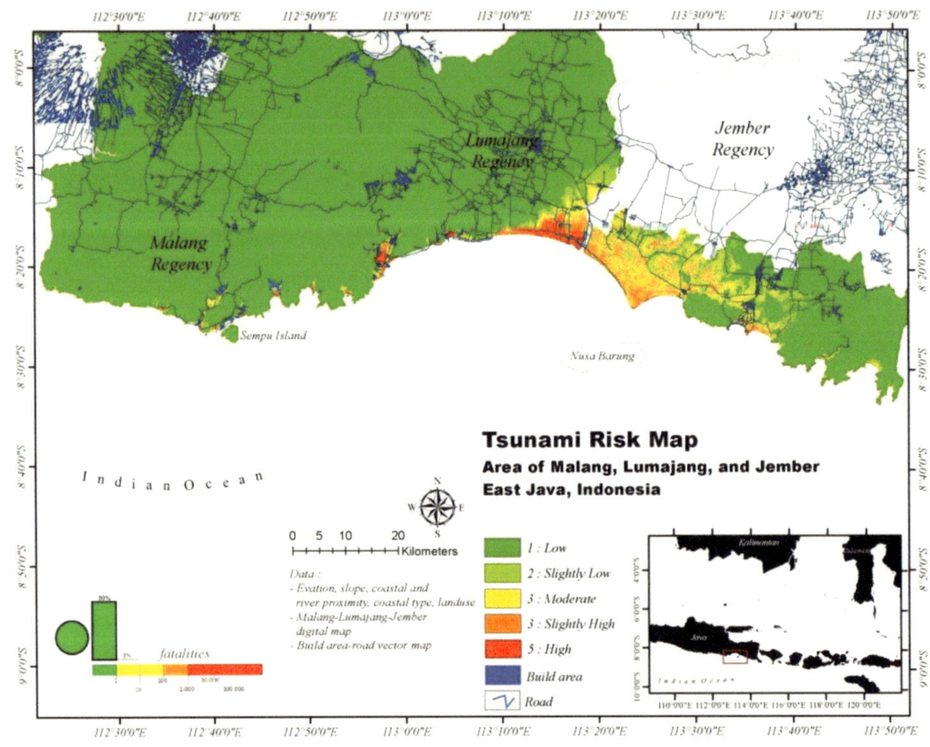

Figure 9.
Tsunami risk map.

out after the 2011 Japan tsunami and found that coastal vegetation is also an important feature to reduce tsunami wave [30]. The high density of mangrove along the coastal area has the capacity to minimize the negative impact of a tsunami wave. Dense mangrove forests growing along coastal areas can help reduce the devastating impact of tsunamis and coastal storms by absorbing some of the wave's energy. When the tsunami struck India's southern state of Tamil Nadu on 26 December, for example, coastal areas with dense mangroves (areas of Pichavaram and Muthupet) suffered fewer human casualties and less damage to property compared to areas without mangroves [40].

The use of pair-wise comparison in AHP process helps in the analysis of spatial multicriteria data where all of the parameters were calculated based on their weight factor. The calculation of weight as a result of pair-wise comparison matrix was created from expert judgment, from a person who was selected based on his/her expertise in tsunami hazard, and disaster mitigation. Tsunami vulnerability research in Alexandria applied all parameters in equal weight due to the limitation of knowledge regarding the study area [14].

High tsunami risk areas were mostly found in the coastal area with the sloping coast type. Elevation and slope play an important role in governing the stability of a terrain. Tsunami vulnerability research in Bali, Indonesia, shows the distribution of vulnerability is not uniform and physically it is highly influenced by coastal proximity, elevation, and slope [29, 37]. Tsunami risk map that described here is based on the integrated approach and provided to the people in the near future due to less information about tsunami risk in the study area.

4. Conclusions

Tsunami risk can be assessed using the application of spatial multicriteria analysis followed by weighted cell-based processing in terms of GIS. DEM data were applied as basic data for creating the parameter of tsunami vulnerability. The result performed here can be used for the evacuation and reconstruction plan due to the tsunami disaster. Also, this will be important basic information in determining both evacuation building and the evacuation route in the coastal areas. Moreover, the final target of tsunami risk mapping is to reduce the effect of the tsunami to the coastal areas where the population is dense by generating a good mitigation plan. The integrated approach of raster weighted overlay of all spatial databases of tsunami vulnerability and risk parameters specified the vulnerability and risk area due to the tsunami and defined the possible area that could be affected by the tsunami and the potential inundated area. The weight of each parameter was calculated by pair-wise comparison matrix from the construction of expert judgment, in which every parameter was weighted not equally.

The overlay processing of tsunami risk map and existing land use, also the distribution of infrastructure and main public facilities, will define the priority area that needs first to be evacuated when tsunami happens. The result of weighted overlay illustrated that high land use of tsunami vulnerability and tsunami risk mostly in the class of urban area in which it is describe the high density of population. Forest area was indicated in the low class of tsunami risk. The green belt mitigation is one of the projects to construct the distribution of coastal vegetation (mangrove) and set it as the barrier zone to reduce the energy of tsunami wave when comes to coastal area and to minimize the negative impact of the wave to the coastal area. More parameters of physical and social vulnerability to tsunami disaster are needed to produce more detailed result.

Acknowledgements

The authors are thankful to Ministry of Research, Technology and Higher Education Indonesia for financial support through the program of applied research funding competition, METI and NASA for the Aster GDEM products, Japan Aerospace Exploration Agency (JAXA) for the ALOS images, Intermap for providing NEXTMap World 10 digital elevation model (DEM), Geospatial Authority of Indonesia for providing the basic map of the study area, and U.S. Geological Survey for seismic data. The authors also thank Laboratory of Disaster Prevention System, Faculty of Engineering, Yamaguchi University, Japan, and Laboratory of Marine Resources Exploration, University of Brawijaya, Indonesia. The Marine Resources Exploration and Management (MEXMA) Research Group contributed to the survey and data collection.

Conflict of interest

The authors declare no conflict of interest.

Author details

Abu Bakar Sambah[1*] and Fusanori Miura[2]

1 Faculty of Fisheries and Marine Science, University of Brawijaya, Malang, Indonesia

2 Environmental Safety Science and Engineering, Graduate School of Science and Engineering, Yamaguchi University, Ube-shi, Japan

*Address all correspondence to: absambah@ub.ac.id

References

[1] National Geophysical Data Center/ World Data Service (NGDC/WDS). Global Historical Tsunami Database. U.S: National Geophysical Data Center. NOAA; 2014. DOI: 10.7289/V5PN93H7 [Accessed: 4 March, 2014]

[2] Tsunami Generation. Available from: http://pubs.usgs.gov/circ/c1187/

[3] Murata S, Imamura F, Katoh K, Kawata Y, Takahashi S, Takayama T. Tsunami; to Survive from Tsunami. Advanced Series on Ocean Engineering. 2009;**32**:177-239. DOI: 10.1142/7345

[4] Kimura S. Journal of Computer Science. 1987;**1**(2):23-49

[5] Islam MR. Conference proceedings. In: Proceedings of the 2nd International Conference on GEOMATE; 2011. pp. 8-13

[6] UN-ISDR (United Nations International Strategy for Disaster Reduction). 2009. Available from: https://www.unisdr.org/we/inform/terminology

[7] Sambah AB, Miura F, Guntur, Fuad. Integrated satellite remote sensing and geospatial analysis for tsunami risk assessment. International Journal of GEOMATE. 2018;**14**(44):96-101. DOI: 10.21660/2018.44.7127

[8] Sambah AB, Miura F. Spatial data analysis and remote sensing for observing tsunami-inundated areas. International Journal of Remote Sensing. 2016;**37**(9):2047-2065. DOI: 10.1080/01431161.2015.1136450

[9] Karen E, Joyce KC, Wright S, Samsonov V, Ambrosia VG. Remote sensing and the disaster management cycle. In: Jedlovec G, editor. Advances in Geoscience and Remote Sensing. Croatia: InTech; 2009. DOI: 10.5772/8341

[10] Yamazaki F, Kouchi K, Matsuoka M. Tsunami damage detection using moderate-resolution satellite imagery. In: Proceedings of the 8th U.S. National Conference on Earthquake Engineering. San Francisco, CA; 2006. pp. 18-22

[11] Yamazaki F, Matsuoka M. Remote sensing technology in post-disaster damage assessment. Journal of Earthquakes and Tsunamis. World Scientific Publishing Company. 2007;**1**(3):193-210. DOI: 10.1142/S1793431107000122

[12] Carver SJ. Integrating multi-criteria evaluation with geographical information systems. International Journal of Geographical Information Systems. 1991;**5**(3):321-339. DOI: 10.1080/02693799108927858

[13] Jankowski P. Integrating geographical information systems and multiple criteria decision-making methods. International Journal of Geographical Information Systems. 1995;**9**(3):251-273. DOI: 10.1080/02693799508902036

[14] Eckert S, Jelinek R, Zeug G, Krausmann E. Remote sensing-based assessment of tsunami vulnerability and risk in Alexandria, Egypt. Applied Geography. 2012;**32**(2):714-723

[15] Romer H, Willroth P, Kaiser G, Vafeidis AT, Ludwig R, Sterr H, et al. Potential of remote sensing techniques for tsunami hazard and vulnerability analysis—A case study from Phan-Nga province, Thailand. Natural Hazards and Earth System Sciences. 2012;**12**(6):2103-2126

[16] Mahendra RS, Mohanty PC, Bisoyi H, Kumar TS, Nayak S. Assessment and management of coastal multi-hazard vulnerability along the Cuddalore Villupuram, east coast of India using geospatial techniques. Ocean and Coastal Management. 2011;**54**(4):302-311

[17] Strunz G, Post J, Zosseder K, Wegscheider S, Muck M, Riedlinger T, et al. Tsunami risk assessment in Indonesia. Natural Hazards and Earth System Sciences. 2011;**11**(1):67-82

[18] Sinaga TP, Adhi N, Yang-Won L, Yongcheol S. GIS mapping of tsunami vulnerability: Case study of the Jembrana regency in Bali, Indonesia. KSCE Journal of Civil Engineering. 2011;**15**(3):537-543

[19] Gokon H, Koshimura S. Mapping of building damage of the 2011 Tohoku earthquake tsunami in Miyagi prefecture. Coastal Engineering Journal. 2012;**54**(1):126-138

[20] Mori N, Takahashi T, Yasuda T, Yanagisawa H. Survey of 2011 Tohoku earthquake tsunami inundation and run-up. Geophysical Research Letters. 2011;**38**(7):1-6

[21] Dall'Osso F, Bovio L, Cavalletti A, Immordino F, Gonella M, Gabbianelli G. A novel approach (the CRATER method) for assessing tsunami vulnerability at the regional scale using ASTER imagery. Italian Journal of Remote Sensing. 2010;**42**(2):55-74

[22] Maramai A, Tinti S. The 3 June 1994 Java tsunami: A post-event survey of the coastal effects. Natural Hazards. 1997;**15**:31-49

[23] Storchak DA, Di Giacomo D, Bondár I, Engdahl ER, Harris J, Lee WHK, et al. Public release of the ISC-GEM global instrumental earthquake catalogue (1900-2009). Seismological Research Letters. 2013;**84**(5):810-815

[24] Bouvet M, Chander G, Goryl P, Santer R, Saunier S. Preliminary radiometric calibration assessment of ALOS AVNIR-2. In: Geoscience and Remote Sensing Symposium. Barcelona: IGARSS; 2007. pp. 2673-2676

[25] Sah AK, Sah BP, Honji K, Kubo N, Senthil S. Semi-automated cloud/shadow removal and land cover change detection using satellite imagery. International Archives of the Photogrammetry, Remote Sensing and Spatial Information Sciences. Vol. XXXIX-B7, XXII ISPRS Congress. Melbourne, Australia. 25 August–01 September, 2012. pp. 335-340

[26] Hansen MC, Defries RS, Townshend JRG, Sohlberg R. Global land cover classification at 1km spatial resolution using a classification tree approach. International Journal of Remote Sensing. 2000;**21**:1331-1364. DOI: 10.1080/014311600210209

[27] Han J, Kamper. Data Mining Concept and Techniques. San Francisco, USA: Morgan Kaufmann Publisher; 2011

[28] Friedl MA, Brodley CE. Decision tree classification of land cover from remotely sensed data. Remote Sensing of Environment. 1997;**61**:399-409

[29] Iida K. Magnitude, energy and generation mechanisms of tsunamis and a catalogue of earthquakes associated with tsunamis. In: Proceedings of Tsunami Meeting at the 10th Pacific Science Congress; 1963. pp. 7-18

[30] Van Zuidam RA. Guide to Geomorphologic—Aerial Photographic Interpretation and Mapping. Enschede, The Netherlands: International Institute for Geo-Information Science and Earth Observation; 1983

[31] Sengaji E, Nababan B. Tsunami risk level mapping in Sikka, East Nusa Tenggara. Journal of Tropical Marine Science and Technology. 2009;**1**(1): 48-61 (Indonesia edition)

[32] Smith K. Environmental Hazards-Assessing Risk and Reducing Disaster. 1st ed. London, New York: Routledge; 1992. p. 324

[33] ESRI, Environmental Systems Research Institute, Inc. Analyse Site

Conditions Using Weighted Overlay. New York, USA: ESRI; 2015

[34] Eastman JR, Jin W, Kyem PAK, Toledano J. Raster procedures for multi-criteria/multi-objective decisions, photogrammetric engineering & remote sensing. American Society for Photogrammetry and Remote Sensing. 1995;**61**(5):539-547

[35] Eddy. GIS in disaster management: A case study of tsunami risk mapping in Bali, Indonesia [Masters (Research) Thesis]. Australia: James Cook University; 2006

[36] Niekert D. Introduction to Disaster Risk Reduction. The African Centre for Disaster Studies NWU, Potchefstroom, South Africa: USAID Disaster Risk Reduction Training Course for Southern Africa, Creative Commons Attribution-Share Alike 2.5 South Africa; 2011

[37] Cutter SL, Mitchell JT, Scott MS. Revealing the vulnerability of people and places: A case study of Georgetown country, South Carolina. Annuals of the Association of American Geographers. 2000;**90**:713-737

[38] Papathoma M, Dominey-Howes D. Tsunami vulnerability assessment and its implications for coastal hazard analysis and disaster management planning, Gulf of Corinth, Greece. Natural Hazards and Earth System Sciences. 2003;**3**(6):733 747

[39] Papathoma M, Dominey-Howes D, Zong Y, Smith D. Assessing tsunami vulnerability, an example from Herakleio, Crete. Natural Hazards and Earth System Sciences. 2003;**3**(5):377-389

[40] Zhou H-W. Introduction to seismic data and processing. In: Practical Seismic Data Analysis. Cambridge, United Kingdom: Cambridge University Press; 1991. Available from: http://assets.cambridge.org/97805211/99100/excerpt/9780521199100_excerpt.pdf

Chapter 8

Utilization of Unmanned Aerial Vehicle for Accurate 3D Imaging

Yoichi Kunii

Abstract

In order to acquire geographical data by aerial photogrammetry, many images should be taken from an aerial vehicle. After that, the images are processed with the help of the structure-from-motion (SfM) technique. Multiple neighboring images with a high rate of overlapping should be obtained for high-accuracy measurement. In the event of natural disasters, UAV operation may sometimes involve risk and should be avoided. Therefore, an easy and convenient method of operating the UAVs is needed. Reports exist on some applications of the UAVs with other devices; however, it will be difficult to prepare a number of such devices in emergency. We considered the most suitable condition for image acquisition by using the UAV. Specifically, some of the altitudes and the rate of overlapping were attempted, and accuracies of the 3D measurement were confirmed. Furthermore, we developed a new camera calibration and measurement method that requires only a few images taken in a simple UAV flight. The UAV in this method was flied vertically and the images were taken at a different altitude. As a result, the plane and height accuracy was ±0.093 and ±0.166 m, respectively. These values were of higher accuracy than the results of the usual SfM software.

Keywords: UAV, 3D measurement, camera calibration, overlapping, accuracy

1. Introduction

The demand for the unmanned aerial vehicles (UAVs) is increasing as they find applications in various fields. For example, more accurate geographical data can be acquired by using the UAVs than by using the usual aerial photogrammetry [1]. The UAVs can take high resolution images as they are able to fly at low altitudes [2]. In addition, the UAVs can be used for observation of natural disasters [3, 4] or for surveying the construction sites [5, 6]. Such applications need rapid and low-cost surveying, and the UAVs are well suited for that purpose [7]. In the case of applying this method to the public survey, the manual published by the Geographical Survey Institute of the Ministry of Land, Infrastructure and Transport in Japan prescribes that the overlap ratio between continuous images is 80% or more. Therefore, even in a narrow target area, it should be taking about several dozen sheets. Also, photogrammetry software equipped with SfM (Structure from Motion), which is now mainstream, also supports such a large number of images. However, since the imaging method as described above requires technology and labor for operating the

UAV, there is a concern in terms of cost, such as requiring a dedicated operator when applying in various construction sites or the like.

Therefore, when applying survey by UAV in the landscaping space, with the aim of minimizing the labor of imaging while ensuring adequate measuring accuracy, measuring precision with respect to change in ground level and the number of images taken to be verified. In addition, we tried 3D modeling in urban plaza and hilly terrain by using the obtained results. Furthermore, we developed a new camera calibration and measurement method which requires only a few images taken in a simple UAV flight. The UAV in this method was flied vertically and the images were taken at a different altitude. We compared the measurement accuracy of the proposed method against the SfM method and evaluated the performance of the proposed method by checking the accuracy.

2. Background of UAV photogrammetry

UAV has been developed for military purposes in the United States since the 1950s and has been developed as a small unmanned reconnaissance aircraft around 1970 due to progress of electronic guidance technology and the like. Utilization of UAV in Japan started spreading because it was used since the late 1990s for spraying pesticides; now it is applied in information gathering and surveying at various sites, and its use in media and entertainment is expanding. Among them, aerial photogrammetry is an application field of particular importance. Normal aerial photogrammetry is carried out by a manned aircraft to image the ground above several hundred to several thousand meters from the altitude to the ground, mainly to create a topographic map. On the other hand, since the altitude of the UAV to the ground is as low as several tens to 100 m, it is possible to create a more detailed topographic map than the manned aircraft. Also, since UAV is inexpensive, maneuverable, and easy to operate compared with a manned aircraft, it demonstrates superior ability in capturing terrain during emergencies such as when a disaster occurs. Furthermore, it is expected to be a tool to improve the efficiency of surveying in earthworks and concrete works. Therefore, it can be said that evaluation of measurement accuracy for UAV photogrammetry is required due to such applications.

3. Acquisition of images for evaluation

3.1 UAV devise and test site

The images for checking the accuracy were taken at a UAV test site in Kanagawa, Japan. The UAV test site is managed by the Japan Society for Photogrammetry and Remote Sensing. **Figure 1** shows the entrance to the UAV test site. There are 76 points of circular ground marks that have Japanese national coordinate in the test area of about 5000 m^2, as shown in **Figure 2**. The center coordinates of the ground marks were given by performing the ground survey of the whole site by a total station. This allowed comparing the given coordinates and the results of the UAV photogrammetry and checking the accuracy of the photogrammetry.

Figure 3 shows the UAV "DJI Inspire 1" which was used for taking the images. The camera "FC350" on the Inspire 1 has 4000 $\times$ 2250 pixels and 4 mm focal length.

Figure 1.
Entrance to UAV test site.

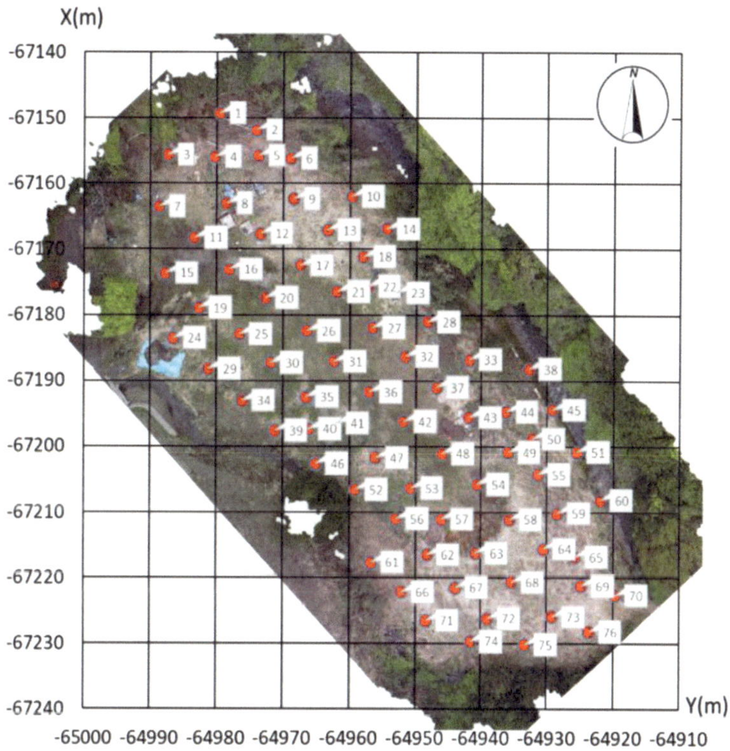

Figure 2.
76 ground marks in the test site.

3.2 Altitude and overlapping rate

The taking the image of the accuracy verification on the test site was carried out on 23 October, 2016. The altitude of the UAV was set as three stages of 40, 60, and 80 m. The taking at each altitude was carried out so that the overlap ratio was 90% and the side lap ratio was 60%. As a result, the number of image acquired at each altitude at the ground level was 135 for 40 m, 57 for 60 m, and 26 for 80 m. **Figure 4** shows samples of images taken at each altitude.

Figure 3.
DJI Inspire 1.

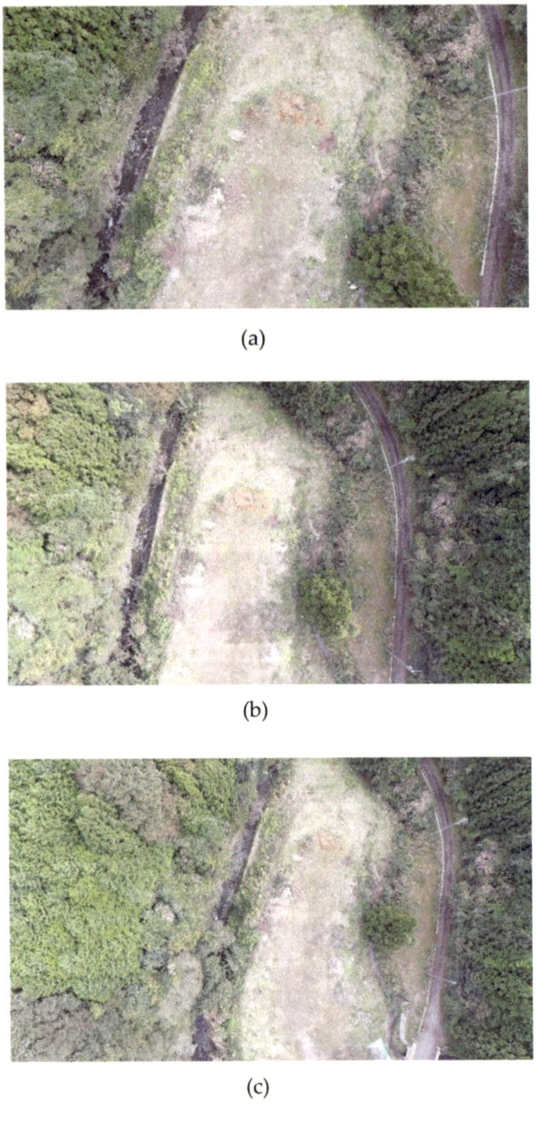

(a)

(b)

(c)

Figure 4.
Sample images at each altitude. (a) 40 m, (b) 60 m, and (c) 80 m.

4. Verification of measurement accuracy

4.1 Details of the verification

3D surveying for each ground mark was carried out by photogrammetry, and accuracy verification was carried out using the image of the test site obtained by such taking image. In the verification of accuracy, in addition to verification by each altitude to ground level, verification is also required when using images with overlap rates of 50, 60, 70, 80, and 90% in the form of thinning took images, respectively. As a result, the number of images in the verification was 135 with the relationship between the ground altitude and the overlap ratio of 40 m and 90%, which was the largest number, and the number of the images was 6 in the case of 80 m and 50%. In the verification, among the 76 points of the ground marks at the test site, 9 points, 13, 17, 25, 33, 42, 47, 70, 75, and 76, were set as control points. On the other hand, the other 67 points were set as verification points, and the accuracy verification of the 3D coordinates obtained for the verification point was decided. For accuracy verification, Agisoft PhotoScan Professional (hereinafter referred to as PhotoScan) which is a general photogrammetry software with SfM was used.

4.2 Results of the verification

In order to verify the accuracy for each condition, root mean square errors (RMSEs) were calculated with the following equation.

$$\sigma_0 = \pm\sqrt{\frac{v_1^2 + v_2^2 + v_3^2 + \cdots + v_n^2}{n-1}} = \pm\sqrt{\frac{[v^2]}{n-1}} \tag{1}$$

where,

σ_0 : RMSE

v : Residual error

n : Number of the data

Table 1 shows the results of accuracy verification carried out as mentioned above. The measurement accuracy shown in the same table is set by setting the value of the ground control point survey at each verification point to the true value and the value of the photogrammetry by the UAV as the measurement value and calculating the standard deviation calculated from the difference value of both at each point. From this result, it can be confirmed that an accuracy of about ± 0.05 m is obtained at any altitude of ground and overlap ratio. According to the precision standard of earthmoving specified by the Ministry of Land, Infrastructure and Transport, if it is within ± 0.1 m, it can be applied to the construction surveying and rock surveying. The results of this verification can be confirmed to satisfy the above-mentioned numerical values in any of the results. On the other hand, it is also possible to apply it to measurement of shape within ± 0.05 m. Regarding this numerical value, at the overlap rate of 90%, any ground altitude is satisfied; however, the result satisfied with 80% or less is mostly at the ground altitude of 40 m. In theory, in the photogrammetry, the altitude of the ground is low, and as the overlap rate becomes higher, the accuracy improves. It is thought that the results according to the situation were obtained.

In addition to the above results, it was confirmed that certain results can be obtained at any ground altitude and overlap rate in this verification. In other words, in the landscaping space where various environments exist in natural space and urban space, the possibility that the method of flight of UAV is limited may be

Overlapped Rate	Altitude	Number of Pictures	Accuracy (RMSE) (m) X, Y	Accuracy (RMSE) (m) Z
50%	40m	27	±0. 024	±0. 047
	60m	12	±0. 015	±0. 046
	80m	6	±0. 016	±0. 054
60%	40m	34	±0. 040	±0. 048
	60m	15	±0. 017	±0. 059
	80m	7	±0. 019	±0. 056
70%	40m	45	±0. 034	±0. 045
	60m	19	±0. 017	±0. 064
	80m	9	±0. 017	±0. 063
80%	40m	68	±0. 035	±0. 047
	60m	29	±0. 021	±0. 054
	80m	13	±0. 019	±0. 046
90%	40m	135	±0. 035	±0. 049
	60m	57	±0. 036	±0. 043
	80m	26	±0. 019	±0. 046

Table 1.
Result of accuracy verification.

considered; however, by this verification, it is possible to fly according to the local situation.

From the above results, it is suggested that the usefulness of surveying in the landscaping space using UAV was suggested, so the case of 3D modeling by UAV conducted at the two survey sites is shown below.

5. Examples of application

5.1 Measurement for plaza

First of all, as an application to UAV's open space in urban space, we decided to do 3D modeling on Yurinoki Plaza at Tokyo University of Agriculture Setagaya Campus (Setagaya, Tokyo) as shown in **Figure 5**. In the Yurinoki Plaza, several trees were planted in a space of about 6000 m^2 covered with lawn. In addition, there are buildings such as research buildings around the open space, and these buildings are also subject to 3D modeling. The images of Yurinoki Plaza were taken by UAV, and a total of 431 images were taken. Of these, 87 images were taken from the UAV in the vertical direction to the ground with an overlap rate of 80% from the altitude of 20 m to the ground and the other 344 aimed the camera in the horizontal direction. **Figure 6** shows samples of the took image.

Figure 5.
Yurinoki Plaza.

Figure 6.
Image from UAV.

The image taken as mentioned above was processed by PhotoScan; 3D point cloud data of each feature captured in the image was generated. Next, a high density point cloud is generated from the obtained point cloud. Compared with the point cloud, the high density point cloud has a high density of point clouds composed of data. Therefore, it seems that texture is attached at the viewpoint from a distance, since it is a set of points to the last, the part where the hole is open as a surface becomes conspicuous because it is a set of points to the last. **Figure 15** shows a target point indicated by the high density point cloud. Finally, texture mapping was performed for the high density point cloud, and a 3D model could be generated as shown in **Figure** 7. In addition, it was confirmed that the area of the plum tree calculated from the created 3D model was 6358.4 m^2, which was almost the same as the area (6357.7 m^2) obtained by ground survey by the total station. In addition, this plaza closed in 2017, and a new research building scheduled for completion in 2020 is being built in this place. In other words, since the results of this report acquired 3D data before closing the open space, it is expected to be utilized as a record of changes in the campus.

5.2 Measurement for mountain area

As an application to the natural space of UAV, it is necessary to perform 3D modeling on the hilly area of about 200,000 m^2 between Tenjinzawa and

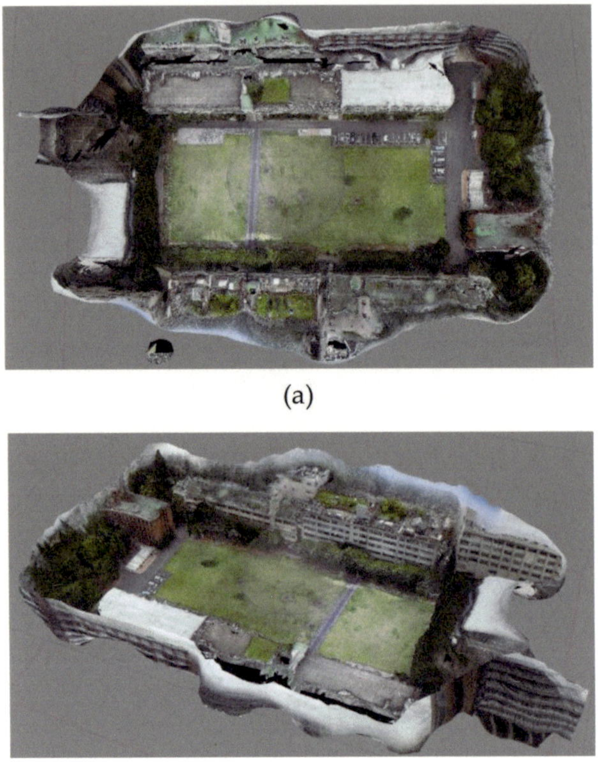

(a)

(b)

Figure 7.
3D model of Yurinoki Plaza. (a) Vertical view, (b) Bird's-eye view.

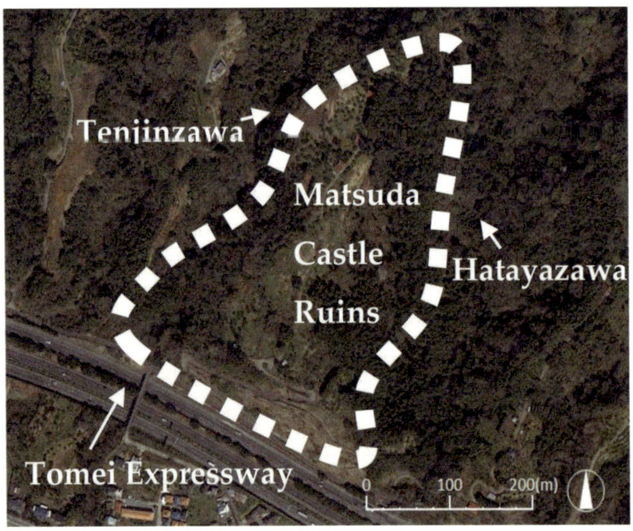

Figure 8.
Matsuda Castle Ruins.

Hatayazawa in Matsuda town, Kanagawa Prefecture, respectively, as shown in **Figure 8**. This area was a place where Matsuda Castle was built in the late Heian era (twelfth century) and is currently managed by Matsuda Town as Matsuda Castle Ruins. The Tomei Expressway passes the southern end of the slope; however, the

Figure 9.
3D modeling for Matsuda Castle Ruins and excavation area.

slope was excavated when constructing this expressway, and there was a circumstance that part of Matsuda Castle site was lost. In this research, the image taken at the Matsuda Castle site by UAV was performed on May 19, 2016, and the image taken with an 80% overlap ratio was secured from the altitude of 70 m. As a result, the number of images taken was 949.

The images taken as mentioned above were processed by PhotoScan. The difference in height from the vicinity of the top of Matsuda Castle to the Tomei Expressway, which was obtained from the created 3D model, is about 66.3 m, which is almost equal to the value (65.5 m) obtained from the Geographical Survey Institute. In addition, the created 3D model also includes the parts excavated by the road construction mentioned above. Therefore, the terrain before excavation clarified by the excavation survey was reproduced as shown in **Figure 9** in complementing the current 3D model. As a result, it is expected that the drilling site by road construction will become visibly apparent, and it will be useful for preservation and management of the future remains.

6. Development of new photogrammetric method

In the above sections, many images should be taken from an aerial vehicle which moves in the horizontal direction and at a fixed altitude [8]. After that, the images are processed with the help of the SfM technique [9]. Multiple neighboring images with a high rate of overlapping should be obtained for high accuracy measurement [10], which calls for labor and cost. In the event of natural disasters, UAV operation may sometimes involve risk [11] and should be avoided. Therefore, an easy and convenient method of operating the UAVs is strongly needed. Reports exist on some applications of the UAVs with other devices [12]; however, it will be difficult to prepare a number of such devices in emergency.

In this research, we developed a method of limiting the movement of UAV only in the vertical direction, using only a small number of images vertically taken at different ground altitudes and performing aerial photogrammetry without using the ground reference point. In addition, in order to evaluate the performance of the developed method, verification was performed by comparing surveying accuracy with general photogrammetry software.

6.1 Acquisition of images for evaluation

The images for checking the accuracy were also taken at a UAV test site, and DJI Inspire 1 also was used for taking the images.

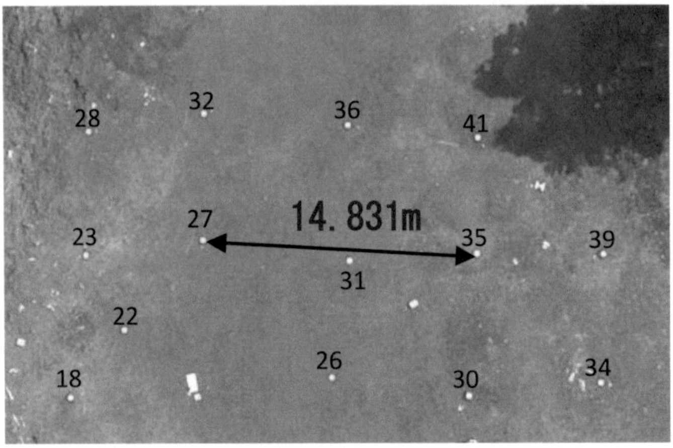

Figure 10.
Given distance between 2 points.

Since the method developed in this research eliminates the ground control point, the obtained 3D coordinates are local coordinates based on arbitrary origin and coordinate axes. Also, to calculate 3D coordinates by this method, it is necessary to give only the distance between two arbitrary points as a known quantity. In this research, we decided to treat the distance (14.831 m) between the airspace signs No. 27 and 35 as shown in **Figure 10** as a known amount.

6.2 Theory of the proposed method

In this research, photogrammetry is carried out by using a plurality of vertical images taken from UAV and acquiring common corresponding points for each image. In general photogrammetry procedures, first of all, after performing orientation processing (camera calibration) to obtain exterior orientation parameters such as shooting points and posture of the camera at the time of shooting and interior orientation parameters such as focal length and lens distortion correction coefficient, 3D surveying of the measurement point will be carried out. However, the method developed in this research is to obtain the optimal solution of each parameter while advancing the camera calibration and the 3D survey at the same time. The details of this method will be described below for each procedure.

6.2.1 Estimation of relative distance

We estimate relative positional relationships with each principal point with respect to a plurality of vertical images taken from the UAV. **Figure 11** schematically shows the situation of the camera at the time of taking each image, and it is assumed to be 1, 2, ... in descending order of altitude to ground. First, the approximate ground altitude for each image was calculated. In this calculation, let the arbitrary two point distance set as described above be a known amount L, let the length on the sensor when L is took on the image be l_1, l_2,... the focal point of the camera when the distance is f, the approximate imaging heights H_1, H_2, ... for each image are obtained by the following equation.

$$H_i = \frac{L}{l_i} f \quad (i = 1, 2, 3, \cdots) \tag{2}$$

where,
H_i : Altitude of pictures (approximate) (m)
L : Given distance (m)
l_i : Given distance on sensors (m)
f : Focal length (approximate) (m)

Therefore, the relative distance between the lowest principal point and the other B_{z1}, B_{z2}, ... and B_{z5} could be calculated by the following equation.

$$B_{zi} = H_i - H_1 \quad (i = 1, 2, 3, \cdots) \tag{3}$$

where,
B_{zi} : Distance between principal points (approximate) (m)
H_i : Altitude of pictures (approximate) (m)

6.2.2 Relative orientation

The relative orientation is to obtain relative took points and postures with respect to a plurality of took images. Generally, relative orientation is often performed only between two images; however, in this study, based on the image No.1 in **Figure 11** as a reference, relative orientation with respect to the other images after image 2. We decided to do all at the same time. In other words, it is assumed that image 1 is taken with no inclination at the origin of the relative coordinates, and the relative point and rotation angle at the time of taking after the image 2 are obtained at the same time. Furthermore, with respect to mutual orientation in this research, the interior orientation parameter of the camera is also set as an unknown quantity as a parameter common to each image, and the orientation is performed at the same time. **Figure 12** shows a coplanar condition that focuses on only No. 1 and 5 images. The principal points of these two images and a common

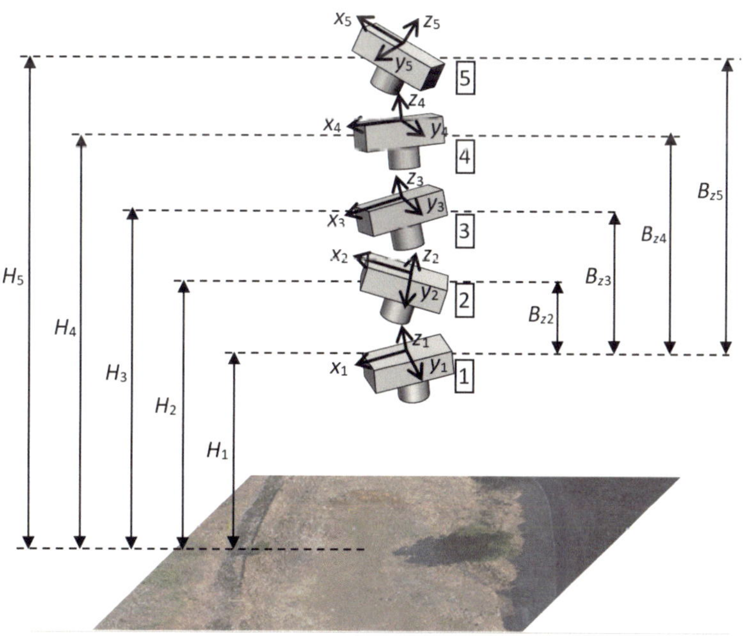

Figure 11.
Positional relation of vertical images.

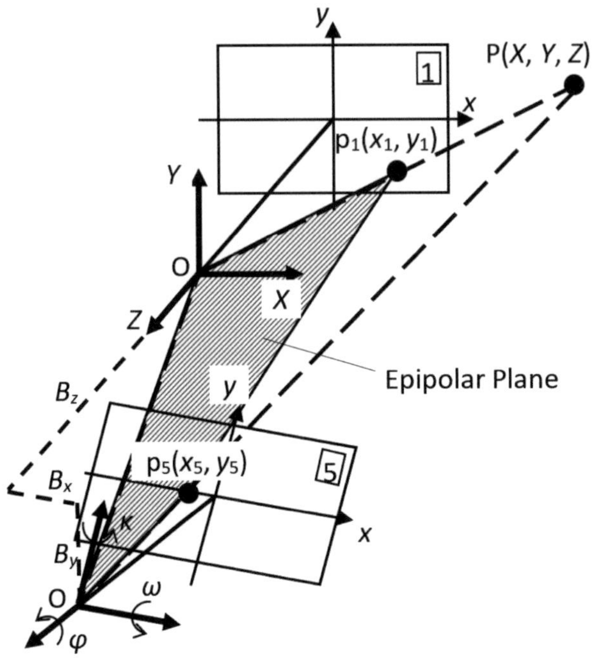

Figure 12.
Coplanarity condition of two vertical images.

point P are set as one plane (epipolar plane). Hereinafter, the details of the present method will be described based on the figure.

Let the principal points of each image be $O_1(0, 0,0)$ and $O_5(B_x, B_y, B_z)$, image points of P be $p_1(x_1, y_1)$ and $p_5(x_5, y_5)$. Then, the relationship of these two images is expressed by the following coplanarity equation.

$$\begin{vmatrix} B_x & B_y & B_z \\ X_1 & Y_1 & Z_1 \\ X_5 & Y_5 & Z_5 \end{vmatrix} = 0$$

where,

$$\begin{pmatrix} X_1 \\ Y_1 \\ Z_1 \end{pmatrix} = \begin{pmatrix} x_1 \\ y_1 \\ -f \end{pmatrix}$$

$$\begin{pmatrix} X_5 \\ Y_5 \\ Z_5 \end{pmatrix} = \mathbf{R} \begin{pmatrix} x_5 \\ y_5 \\ -f \end{pmatrix} + \begin{pmatrix} B_x \\ B_y \\ B_z \end{pmatrix} \quad (4)$$

$$\mathbf{R} = \begin{pmatrix} 1 & 0 & 0 \\ 0 & \cos\omega & -\sin\omega \\ 0 & \sin\omega & \cos\omega \end{pmatrix} \begin{pmatrix} \cos\phi & 0 & \sin\phi \\ 0 & 1 & 0 \\ -\sin\phi & 0 & \cos\phi \end{pmatrix} \begin{pmatrix} \cos\kappa & -\sin\kappa & 0 \\ \sin\kappa & \cos\kappa & 0 \\ 0 & 0 & 1 \end{pmatrix}$$

ω, ϕ, κ : rotation angle of No.5

f : focal length

The relative distance B_{z1} obtained by Eq. (3) is substituted for B_z in Eq. (4). That is, under normal coplanar conditions, B_x is a fixed value; however, in this case, relative orientation is done with B_z as a known quantity. Furthermore, in the mutual orientation in this method, since the interior orientation parameter common to each image is also treated as an unknown quantity, it is necessary to consider the interior orientation parameter with respect to the image coordinates of the image points p_1 and p_5. That is, considering the principal point positions as u_0 and v_0, the scale factors as a_1, a_2, a_3, and a_4, as for the lens distortion, the radiation direction (coefficients: k_1, k_2, and k_3) and the tangential direction (p_1, p_2), (x_i, y_i) $(i = 1, 5)$ in the Eq. (4) is obtained by converting the pixel coordinates (u_i, v_i) (i = 1, 5) obtained from each took image by the following equation.

$$\left.\begin{aligned} x_i &= x'_i + x'_i\left(k_1 r^2 + k_2 r^4 + k_3 r^6\right) + p_1\left(r^2 + 2x'^2_i\right) + 2p_2 x'_i y'_i \\ y_i &= y'_i + y'_i\left(k_1 r^2 + k_2 r^4 + k_3 r^6\right) + p_2\left(r^2 + 2y'_i 2\right) + 2p_1 x'_i y'_i \end{aligned}\right\}$$

$$\begin{aligned} &\text{where,} \\ &k_1, k_2, k_3 : \text{Coefficients of radial distortion} \\ &p_1, p_2 : \text{Coefficients of tangential distortion} \\ &r = \sqrt{x'^2_i + y'^2_i} \\ &u_i = x_p + a_1 x'_i + a_2 y'_i \\ &v_i = y_p + a_3 x' + a_4 y'_i \\ &x'_i, y'_i : \text{Measurement point (mm)} \\ &u_i, v_i : \text{Measurement point (pixel)} \\ &x_p, y_p : \text{Principal point (pixel)} \\ &a_1, a_2, a_3, a_4 : \text{Scale factor} \end{aligned} \tag{5}$$

By sequentially deriving the coplanar conditional expressions from each pair based on image 1, the parameters shown in **Table 2** are unknown quantities in the mutual orientation here. In other words, if one set of corresponding points is obtained between each image, one coplanar condition formula can be obtained, so it is necessary to acquire corresponding points so that a coplanar condition formula exceeding the number of unknown quantities can be obtained. For example, if the number of images is five, the unknown quantity is 10 + 5 × (5 − 1) = 30; however, if 8 or more corresponding points are obtained, the coplanar conditional expression becomes 8 × (5 − 1) = 32 or more, and it is possible to obtain a solution.

Interior Parameters (10)		Relative Orientation Parameters ($5\times(n-1)$)	
Focal Length (mm)	f	Rotation Angle	$\omega_2, \omega_3, \omega_4, \ldots$
Principal Point (pixel)	x_p, y_p		$\varphi_2, \varphi_3, \varphi_4, \ldots$
Scale Factor	a_1, a_2		$\kappa_2, \kappa_3, \kappa_4, \ldots$
Coefficients of Radial Distortion	k_1, k_2, k_3	Camera Position	$B_{x2}, B_{x3}, B_{x4}, \ldots$
Coefficients of Tangential Distortion	p_1, p_2		$B_{y2}, B_{y3}, B_{y4}, \ldots$

Table 2.
Unknown parameters of relative orientation.

6.2.3 Calculation of 3D actual coordinates

Since the relative orientation parameter and the interior orientation parameter for all the images were obtained by the above processing, here, the calculation of the 3D relative coordinates for each measurement point is performed under the collinear condition. The collinear condition is a condition in which the three points, the ground survey point (X, Y, Z), the image point (x, y) on the sensor, and the principal point (X_0, Y_0, Z_0), exist in a straight line. Yes, it is expressed by the following equation as a collinear condition expression.

$$x = -f\frac{a_{11}(X - X_0) + a_{12}(Y - Y_0) + a_{13}(Z - Z_0)}{a_{31}(X - X_0) + a_{32}(Y - Y_0) + a_{33}(Z - Z_0)}$$
$$y = -f\frac{a_{21}(X - X_0) + a_{22}(Y - Y_0) + a_{23}(Z - Z_0)}{a_{31}(X - X_0) + a_{32}(Y - Y_0) + a_{33}(Z - Z_0)}$$

where,

$$\begin{pmatrix} a_{11} & a_{12} & a_{13} \\ a_{21} & a_{22} & a_{23} \\ a_{31} & a_{32} & a_{33} \end{pmatrix} = \begin{pmatrix} 1 & 0 & 0 \\ 0 & \cos\omega & -\sin\omega \\ 0 & \sin\omega & \cos\omega \end{pmatrix} \begin{pmatrix} \cos\phi & 0 & \sin\phi \\ 0 & 1 & 0 \\ -\sin\phi & 0 & \cos\phi \end{pmatrix} \begin{pmatrix} \cos\kappa & -\sin\kappa & 0 \\ \sin\kappa & \cos\kappa & 0 \\ 0 & 0 & 1 \end{pmatrix} \tag{6}$$

That is, since two collinear conditional expressions for one measurement point are obtained for each image, if there are two or more images, it is possible to obtain 3D relative coordinates by $2 \times 2 = 4$ or more collinear conditional expressions. It is possible to solve the three unknown quantities. As a result, 3D relative coordinates for all measurement points are obtained.

Further, all of the obtained 3D relative coordinates are converted into the coordinates of the real scale by the length given as the known amount as shown in **Figure 10**. That is, from the ratio between the actual length and the length on the sensor between points known as known amounts, the 3D relative coordinates for all the measurement points are converted to the coordinates on the real scale. When converting to real scale coordinates, it is necessary to set the coordinate origin and coordinate axes.

6.2.4 Absolute orientation

Since the 3D coordinates on the real scale with respect to all the measurement points are obtained by the above processing, here, the interior orientation parameter common to each image and each exterior orientation parameter are determined by absolute orientation. In other words, in this orientation, all collinear conditional expressions are derived with all measurement points from which 3D coordinates are obtained as ground reference points, and the interior orientation parameter shown in **Table 2** and the took points and attitude angles for each image. All exterior orientation parameters are to be obtained at the same time. As a result, absolute orientation for each image is completed.

6.2.5 Final orientation

The orientation parameters for every camera and the absolute 3D coordinates for every measurement point were acquired by the procedure described above. However, errors in estimation of the absolute 3D coordinates are possible due to conversion from the relative coordinates if using only one given distance. Therefore, as the final stage of this measurement process, all three orientation parameters for all the orientation parameters and all the measurement points are regarded as unknown quantities and the final line. The orientation process shall be carried out.

6.3 Checking accuracy

In order to evaluate the performance of the proposed method, image taking was carried out at the UAV test site and the measurement accuracy verification was carried out. The images were taken by UAV in the vertical direction from the center of the UAV test site. Moreover, every 5 m in the range of the ground altitude of approximately 70–90 m and acquires 5 photos in total as shown in **Figure 13**. In addition, 3D coordinates for 39 points of anti-aircraft signs, which are commonly found in 5 photos, were calculated by this development method and accuracy verification was carried out based on residuals with known coordinates. At that time, as shown in **Figure 14**, the origin is set to No. 27 anti-aircraft marker, the direction of No. 35 is the X axis, the plane formed by these two points and three points is the XY plane, the XY plane is set as the Z axis. In order to applying the proposed orientation method for acquisition of 3D coordinate of these anti-aircraft signs except origin point, the orientation can be performed by using only 2 images. However, in the case of a small number of images, the observation equation and the number of unknown quantities compete with each other, and the convergence state of the calculation by the least squares method becomes unstable. Even in the images taken in this research, trial was done with a small number of sheets; however, it was difficult to stably obtain a convergent solution with 4 or less, so we decided to use all 5 images. **Table 3** shows the results of final orientation for 5 photos. Since the ground altitude in the table is an approximate value obtained by independent positioning with GPS mounted on UAV, a difference of several meters is generated from the Z coordinate in the orientation result.

As shown in **Table 4**, the accuracy verification results showed that the mean square error was within ±0.200 m for both plane and height. When this precision is applied to the surveying accuracy at the earthmoving site, it is considered that the 3D point group within the position accuracy of 0.20 m can be applied to partial payment measurement, and it was recognized that it can be applied as a simple method for earthwork.

Meanwhile, as a comparison target, measurement accuracy was also calculated by general photogrammetry software. The software used is PhotoScan. In this research, PhotoScan also captured the 5 images shown in **Figure 13** and calculated

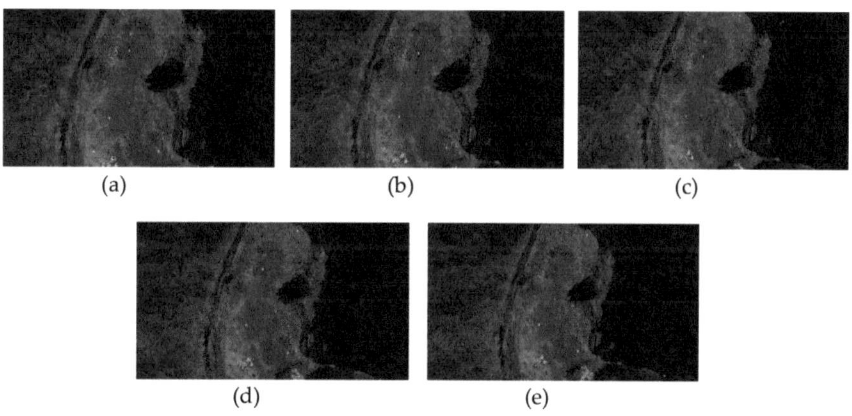

Figure 13.
Vertical images for checking accuracy. (a) 70 m, (b) 75 m, (c) 80 m, (d) 85 m, and (e) 90 m.

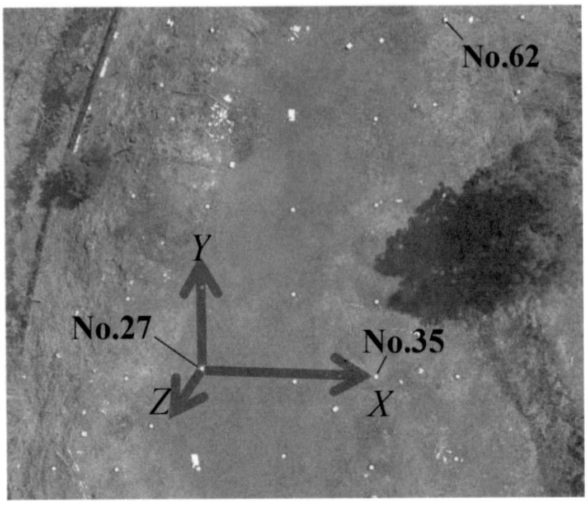

Figure 14.
Local coordinate system.

	Altitude	70 m	75 m	80 m	85 m	90 m
Camera position	X_0(m)	13.648	13.823	14.302	14.180	14.325
	Y_0(m)	-0.619	-1.052	-1.297	-1.528	-1.958
	Z_0(m)	74.061	78.676	83.806	89.035	94.142
Rotation angle	ω	6′ 58″	6′ 43″	6′ 35″	6′ 33″	6′ 28″
	φ	1′ 12″	1′ 15″	1′ 10″	1′ 09″	0′ 59″
	κ	3′ 15″	3′ 00″	3′ 00″	3′ 14″	2′ 58″

Table 3.
Results of final orientation.

Measurement method		RMSE (m) X	Y	Z
Standard accuracy for 5 vertical images		±0.040	±0.040	±0.233
Proposal method	5 vertical images, no GCPs	±0.096	±0.160	±0.166
PhotoScan	5 vertical images, 3 GCPs	±4.808	±3.437	±10.109
	5 vertical images, 39 GCPs	±0.094	±0.098	±0.574
	57 horizontal images, 9 GCPs	±0.038	±0.030	±0.048

Table 4.
Results of checking accuracy.

the 3D coordinates and measurement accuracy for anti-aircraft signs. At that time, we tried two patterns, one with only the No. 27, 35, and 62 as the reference point and one with the 39 points as the reference point. In addition, we also decided to compare it with the case where shooting was performed by a general method as photogrammetry. In other words, with the altitude of the ground set constant at approximately 70 m, the UAV was made to fly in parallel, a total of 57 images were taken to cover the entire test site with securing an overlap rate of 80%, and the image was taken into PhotoScan. The 3D coordinates and measurement accuracy in the case were also calculated. At that time, we decided to use the same number of 9 points as the standard photogrammetry. Also, as an index for evaluating each measurement accuracy obtained above, the standard accuracy generally used in photogrammetry was calculated by the following equation [13].

$$
\begin{aligned}
&\sigma_x = \sigma_y = \frac{H}{f}\sigma_p, \quad \sigma_z = \sqrt{2}\frac{H}{f}\frac{H}{B}\sigma_p \\
&\text{where,} \\
&\sigma_x, \sigma_y, \sigma_z : \text{Standard error for each axis (m)} \\
&H : \text{Altitude (m)} \\
&f : \text{Focal length (m)} \\
&B : \text{Base line (m)} \\
&\sigma_p : \text{Pointing accuracy (m)}
\end{aligned}
\tag{7}
$$

In the above equation, since five vertical took images are used in this study, H in Eq. (7) is the average value (83.944 m) of ground altitude after orientation for 5 images, B is 5. The standard accuracy was calculated using the distance (20.137 m) between the two most distant images. As for the reading accuracy, as in the general photogrammetry, one pixel was used, and the pixel was converted into the length on the sensor of the camera and was used.

As a result, in the case of using only 5 vertical images, the plan accuracy was lower than the standard accuracy for both the proposed method and PhotoScan; however, for the height accuracy only the proposed method exceeded the standard accuracy. In other words, it was confirmed that the proposed method can obtain the accuracy equivalent to that of ordinary photogrammetry, especially in the height direction, although the imaging method is simple and the ground reference point is unnecessary. On the other hand, when images taken by general parallel imaging were processed by PhotoScan, the accuracy was high enough to be applicable to the volume control of the earthworks. From the above results, it is necessary to select the shooting method by UAV according to the situation; however, it can be said that the proposed method is useful for grasping the situation of the site easily in a short time.

6.4 Consideration of the results

Figure 15 shows distribution of residuals of X and Y coordinates with respect to 39 points in an arrow direction. From the figure, within the range of about 10–20 m from the origin, the residuals at most verification points are within ±0.04 m, which is equivalent to the standard accuracy, but No. 45, 50 and 51, 55, 61, and 66, the residual is around ±0.2 m, and it can be confirmed that the accuracy deteriorates. Also, in the same figure, the distribution of the verification point positions can be confirmed to be relatively wide, ranging from about 40 m in the X direction to 50–60 m in the Y direction. In other words, in this verification, it is speculated that the verification point where the Y coordinate is far from the origin is due to the

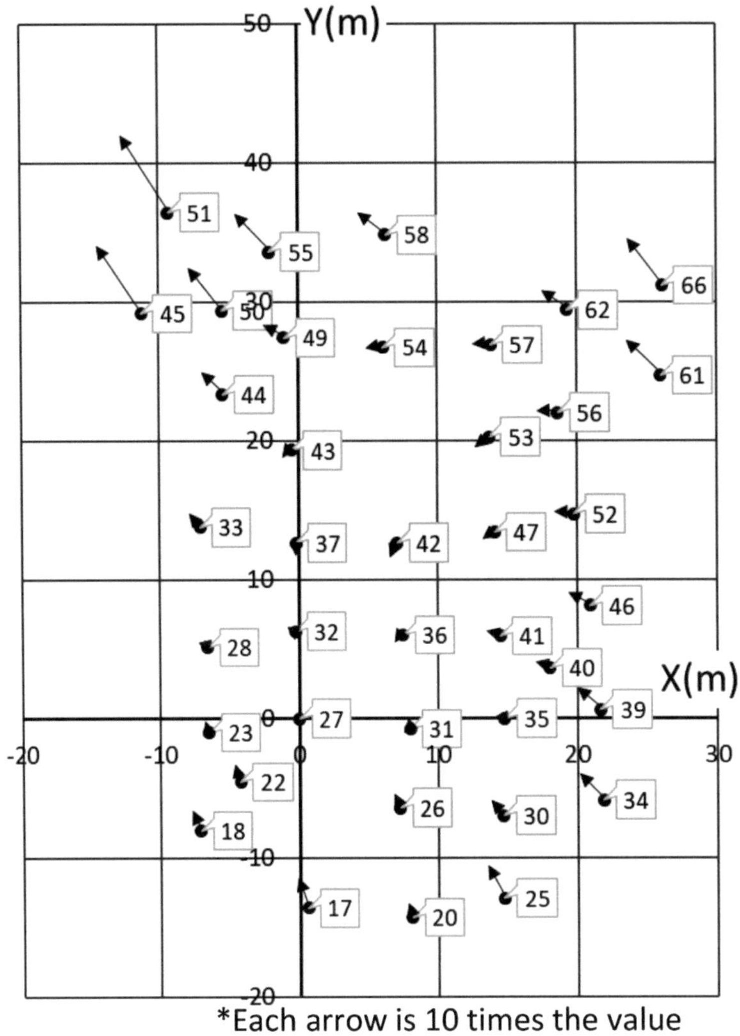

Figure 15.
Error distribution of the proposed method.

decrease in the Y coordinate accuracy in the proposed method. From these results, it is considered preferable to set the origin as close to the measurement object as possible when applying this method in the field.

On the other hand, in order to confirm the utility for 3D measurement by the UAV, measurement accuracy of this result was compared with measurement accuracy by satellite image [14] and aerial image [15]. As a result, the RMSEs of measurement by using the satellite image were ±0.3 to 1.0 m, and the case of the aerial image were ±0.1 to 0.5 m. Such results were dependent on several number of GCPs. Therefore, it can be said the UAV is utilized for accurate measurement in a limited area.

7. Conclusions

In this research, we developed a method of using an image taken vertically from UAV and performing aerial photogrammetry without using the ground reference point. In addition, in order to evaluate the performance of the developed method, verification was performed by comparing surveying accuracy with general photogrammetry software. As a result, since the developed method uses only a small number of vertical took images, it is presumed that the imaging effort can be reduced as compared with the usual method. Also, since the ground reference point is unnecessary, preparation for imaging is unnecessary.

On the other hand, the accuracy verification was performed by comparing with the accuracy of the ground survey by the total station; however, it is inferior in the case of using the general imaging method and software, it was confirmed that the measurement with accuracy of. Specifically, in the general method, it is about ±0.040 m, whereas in the proposed method, it is about ±0.100 m. Also, since the shooting method simply shoots UAV in the vertical direction and shoots several images, it is possible to drastically reduce the time and labor involved in shooting. From these facts, it is expected that the present development method will be used for surveying the current conditions at the earthmoving site and grasping the damage situation at the time of a disaster.

As a future task, we need to consider means for further improving accuracy. In particular, since it is confirmed that the accuracy of this method decreases with respect to a point away from the origin, it is desirable to stabilize the accuracy with respect to the position of the measurement point. Specific countermeasures include verifying the optimum number of photos according to the situation, verifying the optimum altitude difference between the photos, and using GNSS (GPS) positioning information at UAV flight. In this study, 3D coordinates are obtained as local coordinates without using the ground reference point; however, it is necessary to continue discussion on a method for efficiently obtaining global coordinates such as planar rectangular coordinates.

Author details

Yoichi Kunii
Department of Landscape Architecture Science, Tokyo University of Agriculture, Tokyo, Japan

*Address all correspondence to: y3kunii@nodai.ac.jp

References

[1] Valavanis K, Vachtsevanos G, editors. Handbook of Unmanned Aerial Vehicles. Netherlands: Springer; 2015. DOI: 10.1007/978-90-481-9707-1

[2] Beaudoin L, Avanthey L, Gademer A, Roux M, Rudant J. Dedicated payloads for low altitude remote sensing in natural environments. In: International Archives of the Photogrammetry, Remote Sensing and Spatial Information Sciences. Vol. XL-3/W3. La Grande Motte, France; 2015. pp. 405-410

[3] Galarreta J, Kerle N, Gerke M. UAV-based urban structural damage assessment using object-based image analysis and semantic reasoning. Natural Hazards and Earth System Sciences. 2015;**15**:1087-1101. DOI: 10.5194/nhess-15-1087-2015

[4] Li M, Li D, Fanb D. A study on automatic UAV image mosaic method for paroxysmal disaster. In: The International Archives of the Photogrammetry, Remote Sensing and Spatial Information Sciences. Vol. XXXIX-B6. Melbourne, Australia; 2012. pp. 123-128

[5] Barazzetti L, Brumana R, Oreni D, Previtali M, Roncoroni F. True-orthophoto generation from UAV images: Implementation of a combined photogrammetric and computer vision approach. In: ISPRS Annals of the Photogrammetry, Remote Sensing and Spatial Information Sciences. Vol. II-5. Riva del Garda, Italy; 2014. pp. 77-81

[6] Feifei X, Zongjian L, Dezhu G, Huad L. Study on construction of 3D building based on UAV images. In: The International Archives of the Photogrammetry, Remote Sensing and Spatial Information Sciences. Vol. XXXIX-B1. Melbourne, Australia; 2012. pp. 469-473

[7] Tanzi T, Chandra M, Isnard J, Camara D, Sebastien O, Harivelo F. Towards "drone-borne" disaster management: Future application scenarios. In: ISPRS Annals of the Photogrammetry, Remote Sensing and Spatial Information Sciences. Vol. III-8. Prague, Czech Republic; 2016. pp. 181-189

[8] Amrullah C, Suwardhi D, Meilano I. Product accuracy effect of oblique and vertical non-metric digital camera utilization in UAV-photogrammetry to determine fault plane. In: International Archives of the Photogrammetry, Remote Sensing and Spatial Information Sciences. Vol. XLI-B1. Prague, Czech Republic; 2016. pp. 41-48

[9] Westoby J, Brasinton J, Glasser F, Hambrey J, Reynolds M. 'Structure-from-motion' photogrammetry: A low-cost, effective tool for geoscience applications. Geomorphology. 2012;**179**: 300-314

[10] Bagheri O, Ghodsian M, Saadatseresht M. Reach scale application of UAV + SfM method in shallow rivers hyperspatial bathymetry. In: The International Archives of the Photogrammetry, Remote Sensing and Spatial Information Sciences. Vol. XL-1/W5. Kish Island, Iran; 2015. pp. 77-81

[11] Longhitano G, Quintanilha J. Rapid acquisition of environmental information after accidents with hazardous cargo through remote sensing by UAV. In: The International Archives of the Photogrammetry, Remote Sensing and Spatial Information Sciences. Vol. XL-1/W1. Hannover, Germany; 2013. pp. 201-205

[12] Persad R, Armenakis C. Co-registration of DSMs generated by UAV and terrestrial laser scanning systems.

In: The International Archives of the Photogrammetry, Remote Sensing and Spatial Information Sciences. Vol. XLI-B1. Prague, Czech Republic; 2016. pp. 985-990

[13] Yanagi H, Chikatsu H. Performance evaluation of 3D modeling software for UAV photogrammetry. In: The International Archives of the Photogrammetry, Remote Sensing and Spatial Information Sciences. Vol. XLI-B5. Prague, Czech Republic; 2016. pp. 147-152

[14] Rupnika E, Deseillignya MP, Delormeb A, Klingerb Y. Refined satellite image orientation in the free open-source photogrammetric tools Apero/Micmac. In: ISPRS Annals of the Photogrammetry, Remote Sensing and Spatial Information Sciences. Vol. III-1. Czech Republic; 2016. pp. 83-90. DOI: 10.5194/isprsannals-III-1-83-2016

[15] Jung J, Bang K, Sohn G, Armenakis C. Matching aerial images to 3D building models based on context-based geometric hashing. In: ISPRS Annals of the Photogrammetry, Remote Sensing and Spatial Information Sciences. Vol. III-1. Czech Republic; 2016. pp. 17-23. DOI: 10.5194/isprsannals-III-1-17-2016